T0233825

SpringerBriefs in Applied Sciences and Technology

Computational Intelligence

Series editor

Janusz Kacprzyk, Warsaw, Poland

About this Series

The series "Studies in Computational Intelligence" (SCI) publishes new developments and advances in the various areas of computational intelligence—quickly and with a high quality. The intent is to cover the theory, applications, and design methods of computational intelligence, as embedded in the fields of engineering, computer science, physics and life sciences, as well as the methodologies behind them. The series contains monographs, lecture notes and edited volumes in computational intelligence spanning the areas of neural networks, connectionist systems, genetic algorithms, evolutionary computation, artificial intelligence, cellular automata, self-organizing systems, soft computing, fuzzy systems, and hybrid intelligent systems. Of particular value to both the contributors and the readership are the short publication timeframe and the world-wide distribution, which enable both wide and rapid dissemination of research output.

More information about this series at http://www.springer.com/series/10618

Fernando Gaxiola · Patricia Melin
Fevrier Valdez

New Backpropagation Algorithm with Type-2 Fuzzy Weights for Neural Networks

 Springer

Fernando Gaxiola
Division of Graduate Studies
Tijuana Institute of Technology
Tijuana, Baja California
Mexico

Fevrier Valdez
Division of Graduate Studies
Tijuana Institute of Technology
Tijuana, Baja California
Mexico

Patricia Melin
Division of Graduate Studies
Tijuana Institute of Technology
Tijuana, Baja California
Mexico

ISSN 2191-530X   ISSN 2191-5318   (electronic)
SpringerBriefs in Applied Sciences and Technology
ISBN 978-3-319-34086-9   ISBN 978-3-319-34087-6   (eBook)
DOI 10.1007/978-3-319-34087-6

Library of Congress Control Number: 2016939928

© The Author(s) 2016
This work is subject to copyright. All rights are reserved by the Publisher, whether the whole or part of the material is concerned, specifically the rights of translation, reprinting, reuse of illustrations, recitation, broadcasting, reproduction on microfilms or in any other physical way, and transmission or information storage and retrieval, electronic adaptation, computer software, or by similar or dissimilar methodology now known or hereafter developed.
The use of general descriptive names, registered names, trademarks, service marks, etc. in this publication does not imply, even in the absence of a specific statement, that such names are exempt from the relevant protective laws and regulations and therefore free for general use.
The publisher, the authors and the editors are safe to assume that the advice and information in this book are believed to be true and accurate at the date of publication. Neither the publisher nor the authors or the editors give a warranty, express or implied, with respect to the material contained herein or for any errors or omissions that may have been made.

Printed on acid-free paper

This Springer imprint is published by Springer Nature
The registered company is Springer International Publishing AG Switzerland

Preface

In this book, a neural network learning method with type-2 fuzzy weight adjustment is proposed. The mathematical analysis of the proposed learning method architecture and the adaptation of type-2 fuzzy weights are presented. The proposed method is based on research of recent methods that handle weight adaptation and especially fuzzy weights.

The internal operation of the neuron is changed to work with two internal calculations for the activation function to obtain two results as outputs of the proposed method. Simulation results and a comparative study among monolithic neural networks, neural network with type-1 fuzzy weights, and neural network with type-2 fuzzy weights are presented to illustrate the advantages of the proposed method.

The proposed approach is based on recent methods that handle adaptation of weights using fuzzy logic of type-1 and type-2. The proposed approach is applied to the cases of prediction for the Mackey-Glass (for $\tau = 17$) and Dow-Jones time series, and recognition of person with iris biometric measure. In some experiments, noise was applied in different levels to the test data of the Mackey-Glass time series for showing that the type-2 fuzzy backpropagation approach obtains better behavior and tolerance to noise than the other methods.

The optimization algorithms that were used are the genetic algorithm and the particle swarm optimization algorithm and the purpose of applying these methods was to find the optimal type-2 fuzzy inference systems for the neural network with type-2 fuzzy weights that permit to obtain the lowest prediction error.

We describe in Chap. 1 a brief introduction to the potential of the use of type-2 fuzzy weights in the neural networks. We also mention the applications of the proposed methods.

In Chap. 2, some basic theoretical and technical concepts about the areas of computational intelligent, forecasts, and recognition as well as a brief introduction and operation of each are addressed, as all of them are of great importance for the development of this book.

We present in Chap. 3 a clear and accurate explanation of the proposed method of neural network with fuzzy weights; also, the mathematical analysis of the type-1 and type-2 fuzzy weights, and all information used to carry the optimization of the type-2 fuzzy inference systems for the ensemble neural network and monolithic neural network with type-2 fuzzy weights are shown.

In addition, it shows all the architectures of the ensemble neural network and neural network with type-1 and type-2 fuzzy weights implemented; also an explanation of the chromosome with the GA used to optimize the neural network and membership functions for the ensemble neural network with type-2 fuzzy weights are shown. It also shows how type-1 and type-2 fuzzy weights fuzzy Inference System are Implemented for type-2, the membership functions used, and also shows the representation of the chromosome with the GA and the representation of the particles with PSO for the optimization of the membership functions of the type-2 fuzzy systems.

In Chap. 4, we present the results for the proposed method for all study cases: recognition of persons using iris biometric measure with neural network with type-2 fuzzy weights, ensemble neural network with type-2 fuzzy weights and its optimization with GA, neural network with type-1 and type-2 fuzzy weights for triangular and Gaussian membership functions and its optimization with GA, and PSO algorithms for the Mackey-Glass time series with which we work during the development optimization of the book.

In Chap. 5, we presented the conclusions of the proposed method based on the results obtained. In addition, possible future work is outlined.

We end this Preface of the book by giving thanks to all the people who have helped or encouraged us during the writing of this book. First of all, we would like to thank our colleague and friend Prof. Oscar Castillo for always supporting our work and for motivating us to write our research work. We would also like to thank our families for their continuous support during the time that we spent in this project. Of course, we have to thank our institution, Tijuana Institute of Technology, for always supporting our projects. We must thank our supporting agencies, CONACYT y TNM, in our country for their help during this project. Finally, we thank our colleagues working in Soft. Computing, who are too many to mention by name.

Fernando Gaxiola
Patricia Melin
Fevrier Valdez

Contents

Chapter 1
Introduction

In this book we propose an adaptation of weights in the back-propagation algorithm for neural networks using type-2 and type-1 fuzzy inference systems. This proposed approach is different than other ones in the literature, where the adaptation is in the momentum and adaptive learning rate [1–4], or with triangular or trapezoidal fuzzy numbers for the weights [5, 6], and also because the proposed method works with type-1 and type-2 fuzzy weights, which is the main difference with respect to the others methods.

The proposed approach is applied to the recognition of persons using the iris biometric, and time series prediction for the Mackey-Glass and Dow-Jones time series. In this case, the objective is obtaining the recognition of the person and the minimum prediction error for the time series data [7].

This work is based on comparing the performance for the neural network with type-1 fuzzy weights and type-2 fuzzy weights, with the traditional approach of using real numbers for the weights, which is important because the weights affect the performance of the learning process of the neural network and therefore obtaining better results.

This conclusion is based on the application of neural networks of this type, where previous works have shown that the training of neural networks for the same problem initialized with different weights or its adjustment in a different way, but at the end is possible to reach a similar result.

The contribution of the book is the proposed method for type-1 and type-2 fuzzy weight adjustment in back-propagation learning of neural networks for providing them the ability to manage uncertainty in real data. The main idea of the proposed method is that enhancing the back-propagation method with type-1 and type-2 fuzzy logic enables better management of uncertainty in the learning process and with this improved results can be achieved.

The proposed method have tested in two areas, the recognition patterns and time series forecasting; for the recognition patterns we used the iris biometric measure, and for the time series forecasting we used the Mackey-Glass and the Dow-Jones time series.

© The Author(s) 2016
F. Gaxiola et al., *New Backpropagation Algorithm with Type-2 Fuzzy Weights for Neural Networks*, SpringerBriefs in Computational Intelligence,
DOI 10.1007/978-3-319-34087-6_1

We presented two architectures of neural network for tested the proposed method: one, using the monolithic architecture of neural network with type-2 fuzzy weights, and the other one, using ensemble neural network with three neural networks with type-2 fuzzy weights and average integration for the final result.

We optimized the ensemble neural network with genetic algorithm for the number of neurons in the hidden layer of the three neural networks, and the membership functions of the type-2 fuzzy inference systems used in the connections between the input and hidden layer, and between hidden and output layer. In the monolithic neural network with type-2 fuzzy weights, we optimized the membership functions of the type-2 fuzzy inference systems used in the connections between the input and hidden layer, and between hidden and output layer.

Simulation results show that neural networks with the proposed type-2 fuzzy approach have the ability of out performing their type-1 and type-0 counterparts.

The proposed method is different than the adaptive neuro-fuzzy inference system (ANFIS) method, because ANFIS uses the neural network architecture for obtaining the characteristics of the fuzzy systems, and performs the operations based on the calculations of the fuzzy systems; this is different to the proposed method that uses the type-2 fuzzy systems to update the weights used in the neural network for the training (learning) process.

With the application of neural networks we aim for the solution of complex problems, not as a sequence of steps, but as the evolution of computer systems inspired by the human brain, and therefore endowed with some "intelligence" which no are but the combination of simple elements of interconnected processes, which operate in parallel in various styles manage to solve problems related to the recognition of shapes or patterns, prediction, control, optimization, among others.

References

1. Fletcher, R., Reeves, C.M.: Function minimization by conjugate gradients. Comput. J. **7**, 149–154 (1964)
2. Beale, E.M.L.: A Derivation of Conjugate Gradients. In: Lootsma, F.A. (ed.) Numerical methods for nonlinear optimization, pp. 39–43. Academic Press, London (1972)
3. Powell, M.J.D.: Restart procedures for the conjugate gradient method. Math. Program. **12**, 241–254 (1977)
4. Moller, M.F.: A scaled conjugate gradient algorithm for fast supervised learning. Neural Netw. **6**, 525–533 (1993)
5. Ishibuchi, H., Morioka, K., Tanaka, H.: A fuzzy neural network with trapezoid fuzzy weights. In: Fuzzy Systems, vol. 1, IEEE World Congress on Computational Intelligence, pp. 228–233 (1994)
6. Ishibuchi, H., Tanaka, H., Okada, H.: Fuzzy neural networks with fuzzy weights and fuzzy biases. In: IEEE International Conference on Neural Networks, vol. 3, pp. 160–165 (1993)
7. Yadav, V., Srinivasan, D.: A SOM-based hybrid linear-neural model for short term load forecasting. Neurocomputing **74**(17), 2874–2885 (2011)

Chapter 2
Theory and Background

This chapter overviews the background and main definitions and basic concepts, useful to the development of this investigation work.

2.1 Neural Networks

Neural networks are composed of many elements (Artificial Neurons), grouped into layers and are highly interconnected (with the synapses), this structure has several inputs and outputs, which are trained to react (or give values) in a way you want to input stimuli (R values). These systems emulate in some way, the human brain. Neural networks are required to learn to behave (Learning) and someone should be responsible for the teaching or training (Training), based on prior knowledge of the problem environment problem [1].

Artificial neural networks are inspired by the architecture of the biological nervous system, which consists of a large number of relatively simple neurons that work in parallel to facilitate rapid decision-making [2].

An artificial neural network (ANN) is a distributed computing scheme based on the structure of the nervous system of humans. The architecture of a neural network is formed by connecting multiple elementary processors, this being an adaptive system that has an algorithm to adjust their weights (free parameters) to achieve the performance requirements of the problem based on representative samples [3, 4].

The most important property of artificial neural networks is their ability to learn from a training set of patterns, i.e. is able to find a model that fit the data [5, 6].

The artificial neuron consists of several parts (see Fig. 2.1). On one side are the inputs, weights, the summation, and finally the adapter function. The input values are multiplied by weights and added: $\sum x_i w_{ij}$. This function is completed with the addition of a threshold amount i. This threshold has the same effect as an entry with value -1. It serves so that the sum can be shifted left or right of the origin. After

© The Author(s) 2016
F. Gaxiola et al., *New Backpropagation Algorithm with Type-2 Fuzzy Weights for Neural Networks*, SpringerBriefs in Computational Intelligence,
DOI 10.1007/978-3-319-34087-6_2

Fig. 2.1 Scheme of an
artificial neuron

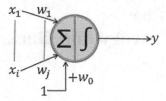

addition, we have the function f applied to the sum, resulting in the final value of
the output, also called y_i [7], obtaining following the equation:

$$y_i = \left(\sum_{i=1}^{n} x_i w_{ij} \right) \tag{2.1}$$

where f may be a nonlinear function with binary output ± 1, a linear function f
$(z) = z$, or as sigmoid logistic function: $f(z) = \frac{1}{1+e^{-z}}$.

A neural network is a system of parallel processors connected as a directed
graph. Schematically each processing element (neuron) of the network is represented as a node. These connections establish a hierarchical structure that is trying
to emulate the physiology of the brain as it looks for new ways of processing to
solve real world problems. What is important in developing the techniques of
Neural Networks (NNs) is if its useful to learn behavior, recognize and apply
relationships between objects and plots of real-world objects themselves. In this
sense, artificial neural networks have been applied to many problems of consid-
erable complexity. Its most important advantage is in solving problems that are too
complex for conventional technologies, problems that have no solution and/or that
the algorithm of the solution is very difficult to find [8].

2.2 Ensemble Neural Networks

An Ensemble Neural Network is a learning paradigm where many neural networks
are jointly used to solve a problem [9]. A Neural network ensemble is a learning
paradigm where a collection of a finite number of neural networks is trained for the
same task [10]. It originates from Hansen and Salamon's work [11], which shows
that the generalization ability of a neural network system can be significantly
improved through a number of neural networks in ensemble, i.e. training many
neural networks and then combining their predictions. Since this technology
behaves remarkably well, recently it has become a very hot topic in both neural
networks and machine learning communities [12], and has already been success-
fully applied to diverse areas such as face recognition [13, 14], optical character
recognition [15], scientific image analysis [16], medical diagnosis [17, 18], seismic
signals classification [19], etc.

In general, a neural network ensemble is constructed in two steps, i.e. training a number of component neural networks and then combining the component predictions.

There are also many other approaches for training the component neural networks. Examples are as follows. Hampshire and Waibel [20] and Wei and Cheng [21] utilize different objective functions to train distinct component neural networks. Cherkauer [16] trains component networks with different number of hidden layers. Maclin and Shavlik [22] initialize component networks at different points in the weight space. Liu et al. [23] and Soltani [24] employ cross-validation to create component networks. Krogh and Vedelsby [25] and Opitz and Shavlik [26] exploit genetic algorithm to train diverse knowledge based component networks. Yao and Liu [27] regard all the individuals in an evolved population of neural networks as component networks [28].

2.3 Type-2 Fuzzy Systems

The basics of fuzzy logic do not change from type-1 to type-2 fuzzy sets, and in general, will not change for any type-n [29–33]. A higher-type number just indicates a higher "degree of fuzziness". Since a higher type changes the nature of the membership functions, the operations that depend on the membership functions change; however, the basic principles of fuzzy logic are independent of the nature of membership functions and hence, do not change. Rules of inference like Generalized Modus Ponens or Generalized Modus Tollens continue to apply.

The structure of the type-2 fuzzy rules is the same as for the type-1 case because the distinction between type-2 and type-1 is associated with the nature of the membership functions. Hence, the only difference is that now some or all the sets involved in the rules are of type-2 [34, 35].

As an example: Consider the case of a fuzzy set characterized by a Gaussian membership function with mean m and a standard deviation that can take values in $[\sigma_1, \sigma_2]$ i.e.,

$$\mu(x) = \exp\left\{-1/2\left[\frac{x - m}{\sigma}\right]^2\right\}; \ \sigma \epsilon [\sigma_1, \sigma_2] \tag{2.2}$$

Corresponding to each value of σ we will get a different membership curve (as shown in Fig. 2.1). So, the membership grade of any particular x (except x = m) can take any of a number of possible values depending upon the value of σ i.e., the membership grade is not a crisp number, it is a fuzzy set. Figure 2.2 shows the domain of the fuzzy set associated with x = 0.7; however, the membership function associated with this fuzzy set is not shown in Fig. 2.2.

Fig. 2.2 A type-2 fuzzy set representing a type-1 fuzzy set with uncertain standard deviation

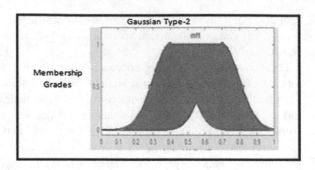

2.3.1 Gaussian Type-2 Fuzzy Set

A Gaussian type-2 fuzzy set is one in which the membership grade of every domain point is a Gaussian type-1 set contained in [0, 1].

2.3.2 Interval Type-2 Fuzzy Set

An interval type-2 fuzzy set is one in which the membership grade of every domain point is a crisp set whose domain is some interval contained in [0, 1].

2.3.3 Footprint of Uncertainty

Uncertainty in the primary memberships of a type-2 fuzzy set, Ã, consists of a bounded region that we call the "footprint of uncertainty" (FOU). Mathematically, it is the union of all primary membership functions [36, 37].

2.3.4 Upper and Lower Membership Functions

An "upper membership function" and a "lower membership functions" are two type-1 membership functions that are bounds for the FOU of a type-2 fuzzy set Ã. The upper membership function is associated with the upper bound of the FOU(Ã). The lower membership function is associated with the lower bound of the FOU(Ã).

2.3.5   Operations of Type-2 Fuzzy Sets

2.3.5.1   Union of Type-2 Fuzzy Sets

The union of $\widetilde{A}_1$ and $\widetilde{A}_2$ is another type-2 fuzzy set, just as the union of type-1 fuzzy sets A_1 and A_2 is another type-1 fuzzy set. More formally, we have the following expression:

$$\widetilde{A}_1 \cup \widetilde{A}_2 = \int_{x \in X} \mu_{\widetilde{A}_1 \cup \widetilde{A}_2}(x)/x \qquad (2.3)$$

2.3.5.2   Intersection of Type-2 Fuzzy Sets

The intersection of $\widetilde{A}_1$ and $\widetilde{A}_2$ is another type-2 fuzzy set, just as the intersection of type-1 fuzzy sets A_1 and A_2 is another type-1 fuzzy set. More formally, we have the following expression:

$$\widetilde{A}_1 \cap \widetilde{A}_2 = \int_{x \in X} \mu_{\widetilde{A}_1 \cap \widetilde{A}_2}(x)/x \qquad (2.4)$$

2.3.5.3   Complement of a Type-2 Fuzzy Set

The complement of a set is another type-2 fuzzy set, just as the complement of type-1 fuzzy set A is another type-1 fuzzy set. More formally we have:

$$\widetilde{A'} = \int_x \mu_{\widetilde{A'}1}(x)/x \qquad (2.5)$$

where the prime denotes complement in the above equation. In this equation $\mu_{\widetilde{A'}1}$ is the secondary membership function.

2.3.6   Type-2 Fuzzy Rules

Consider a type-2 Fuzzy System having r inputs $x_1 \in X_1, \ldots, x_r \in X_r$ and one output $y \in Y$. As in the type-1 case, we can assume that there are M rules; but, in the type-2 case the lth rule has the form:

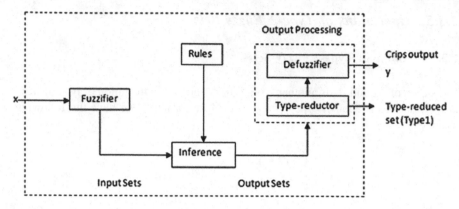

Fig. 2.3 Structure of a type-2 fuzzy logic system

$$R^1 : IF \, x_1 \, is \, \widetilde{A}_1^1 \, and \ldots x_p \, is \, \widetilde{A}_P^1, \, THEN \, y \, is \, \widehat{Y}^1 \quad I = I, \ldots M \qquad (2.6)$$

The rules represent a type-2 relation between the input space $X_1 \times \cdots \times X_r$, and space output set space Y, of the type-2 fuzzy system. In the type-2 fuzzy system (Fig. 2.3), as in the type-1 fuzzy system crisp inputs are first fuzzified into fuzzy input sets than then activate the inference block, which in the present case is associated with type-2 fuzzy sets [38].

2.4   Optimization

Regarding optimization, we have the following situation in mind: there exists a search space V, and a function:

$$g : V \to \mathbb{R} \qquad (2.7)$$

And the problem is to find: ***arg ming(v).***

$$v \in V \qquad (2.8)$$

Here, V is vector of decision variables, and g is the objective function. In this case we have assumed that the problem is one of minimization, but everything we say can of course be applied mutatis mutandis to a maximization problem. Although specified here in an abstract way, this is nonetheless a problem with a huge number of real-world applications.

In many cases the search space is discrete, so that we have the class of combinatorial optimization problems (COPs). When the domain of the g function is continuous, a different approach may well be required, although even here we note that in practice, optimization problems are usually solved using a computer, so that in the final analysis the solutions are represented by strings of binary digits (bits) [39, 40].

There are several optimization techniques for neural networks, some of these are: evolutionary algorithms [41], ant colony optimization [42] and Particle swarm [43].

2.5 Genetic Algorithms

Genetic algorithms were introduced for the first time by a Professor of the University of Michigan, John Holland [44]. A genetic algorithm is a mathematical highly parallel algorithm that transforms a set of mathematical individual objects with regard to the time using operations based on evolution. The Darwinian laws of reproduction and survival of the fittest can be used, and after having appeared of natural form a series of genetic operations that can be used [45, 46]. Each of these mathematical objects is usually a chain of characters (letters or numbers) of fixed length that adjusts to the model of the chains of chromosomes, and one associates to them a certain mathematical function that reflects the fitness.

A Genetic Algorithm (GA) [47] assumes that a possible solution to a problem can be seen as an individual and is represented by a set of parameters. These parameters are the genes of a chromosome, representing the structure of the individual. This chromosome is evaluated by a fitness function, providing a fitness value to each individual, as to the suitability it has to solve the given problem. Through genetic evolution using crossover operations (matching of individuals to produce children) (example: one point crossover, see Fig. 2.4) and mutation that simulates the biological behavior, combined with a selection process (example: single mutation, see Fig. 2.4), the population of individuals is eliminating those with less ability, and tends to improve overall fitness of the population, to produce a solution close to optimal performance for the given problem. John Holland was aware of the importance of natural selection [48], and in the late 60s developed a technique that allowed incorporating it into a computer program. His goal was to get computers to learn for themselves. The technique proposed by Holland was originally called "reproductive plans" [49].

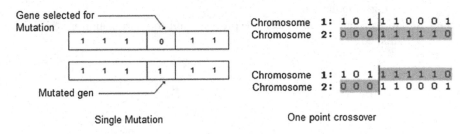

Fig. 2.4 Example of genetic operators

The fitness function determines the best individual where each individual must represent a complete solution to the problem that is being optimized. Therefore the three most important aspects of using genetic algorithms are:

(a) Definition of the fitness function.
(b) Definition and implementation of the genetic representation.
(c) Definition and implementation of the genetic operators.

Therefore, to solve a problem using a GA should take the following steps:

1. Define the chromosome or individual's structure and its representation.
2. Determine the evaluation criteria or adaptation of individuals.
3. Determine the genetic operators to use.
4. Define the stopping criterion of the algorithm.
5. Define a criterion for replacement between consecutive generations.

The basic genetic algorithm is as follows:

Step 1 Represent the problem variable domain as a chromosome of a fixed length; choose the size of a chromosome population N, the crossover probability pc and the mutation probability pm.

Step 2 Define a fitness function to measure the performance, or fitness, of an individual chromosome in the problem domain. The fitness function establishes the basis for selecting chromosomes that will be mated during reproduction.

Step 3 Randomly generate an initial population of chromosomes of size $N = x_1, x_2, \ldots, x_n$.

Step 4 Calculate the fitness of each individual chromosome: $f(x_1), f(x_2), \ldots, f(x_n)$

Step 5 Select a pair of chromosomes for mating from the current population. Parent chromosomes are selected with a probability related to their fitness.

Step 6 Create a pair of offspring chromosomes by applying the genetic operators—crossover and mutation.

Step 7 Place the created offspring chromosomes in the new population.

Step 8 Repeat Step 5 until the size of the new chromosome population becomes equal to the size of the initial population, N.

Step 9 Replace the initial (parent) chromosome population with the new (offspring) population.

Step 10 Go to Step 4, and repeat the process until the termination criterion is satisfied.

2.6 Particle Swarm Optimization

Particle Swarm Optimization (PSO) is a bio-inspired optimization method proposed by Eberhart and Kennedy [50] in 1995. PSO is a search algorithm based on the behavior of biological communities that exhibits individual and social behavior [38, 51], and examples of these communities are groups of birds, schools of fish and swarms of bees [52–54].

A PSO algorithm maintains a swarm of particles, where each particle represents a potential solution. In analogy with the paradigms of evolutionary computation, a swarm is similar to a population, while a particle is similar to an individual. In simple terms, the particles are "flown" through a multidimensional search space, where the position of each particle is adjusted according to its own experience and that of its neighbors. Let x_i denote the position i in the search space at time step t; unless otherwise stated, t denotes discrete time steps. The position of the particle is changed by adding a velocity, $v_i(t)$, to the current position, i.e.

$$x_i(t+1) = x_i(t) + v_i(t+1) \qquad (2.9)$$

With $x_i(0) \sim U(X_{min}, X_{max})$.

It is the velocity vector the one that drives the optimization process, and reflects both the experimental knowledge of the particles and the information exchanged in the vicinity of particles.

For PSO, the particle velocity is calculated as:

$$v_{ij}(t+1) = v_{ij}(t) + c_1 r_{1j}(t)\left[y_{ij}(t) - x_{ij}(t)\right] + c_2 r_{2j}(t)\left[\hat{y}_{ij}(t) - x_{ij}(t)\right] \qquad (2.10)$$

where $v_{ij}(t)$ is the velocity of the particle i in dimension j at time step t, c_1 and c_2 are the positive acceleration constants used to scale the contribution of cognitive and social skills, $\hat{y}_{ij}(t)$ is the best position, respectively, and $r_{1j}(t)$, and $r_{2j}(t) \sim U(0, 1)$ are random values in the range [0, 1].

The best personal position in the next time step $t + 1$ is calculated as:

$$y_i(t+1) = \begin{cases} y_i(t) & \text{if } f(x_i(x_i(t+1)) \geq f\, y_i(t)) \\ x_i(t+1) & \text{if } f(x_i(x_i(t+1)) > f\, y_i(t)) \end{cases} \qquad (2.11)$$

where $f : \mathbb{R}^{nx} \to \mathbb{R}$ is the fitness function, as with EAs, the goal is measuring the fitness function closely corresponding to the optimal solution, for example the objective function quantifies the performance, or the quality of a particle (or solution).

The overall best position $\hat{y}(t)$, at time step t, is defined as:

$$\hat{y}(t)\epsilon\{y_o(t), \ldots, y_s(t)\}f(y(t)) = \min\{f(y_o(t)), \ldots f(y_s(t)),\} \qquad (2.12)$$

where s is the total number of particles in the swarm, more importantly, the above equation defining and establishing $\hat{y}(t)$ the best position is uncovered by either of the particles so far as this is usually calculated as the best personal best position. The global best position can also be selected from the particles of the current swarm, in which case is expressed as:

$$\hat{y}(t) = \min\{f(x_o(t)), \ldots f(x_{ns}(t)),\} \tag{2.13}$$

A few important and interesting modifications in the basic structure of PSO are as follows:

Shi and Eberhart [55] introduced a new parameter called inertia weight (ω) into the velocity vector, Eq. (2.10), after which the equation becomes:

$$v_{ij}(t+1) = \omega * v_{ij}(t) + c_1 r_{1j}(t)\left[y_{ij}(t) - x_{ij}(t)\right] + c_1 r_{1j}(t)\left[\hat{y}_{ij}(t) - x_{ij}(t)\right] \tag{2.14}$$

The inertia weight ω is employed to control the impact of the previous history of velocities on the current velocity, thereby influencing the trade-off between global and local exploration abilities of the particles. It can be a positive constant or even a positive linear or nonlinear function of time. A larger inertia weight encourages global exploration while a smaller inertia weight tends to facilitate local exploration to fine-tune the current search area. Suitable selection of the inertia weight provides a balance between global and local exploration abilities and thus requires less iteration on an average to find the optimum [46]. Initially the inertia weight was kept static during the entire search process for every particle and dimension. However, with the due course of time dynamic inertia weights were introduced.

Clerc [56, 57] introduced a new parameter called constriction factor 'K' that improves PSO's ability to constrain and control velocities. Eberhart and Shi [45] later verified that K, combined with constraints on V_{max}, significantly improved the PSO performance. The value of the constriction factor is calculated as follows:

$$K = \frac{2}{\left|2 - \varphi - \sqrt{\varphi^2 - 4\varphi}\right|}, \varphi = c_1 + c_2, \varphi > 4 \tag{2.15}$$

With the constriction factor K, the PSO equation for computing the velocity is:

$$v_{ij}(t+1) = K\left\{v_{ij}(t) + c_1 r_{1j}(t)\left[y_{ij}(t) - x_{ij}(t)\right] + c_1 r_{1j}(t)\left[\hat{y}_{ij}(t) - x_{ij}(t)\right]\right\} \tag{2.16}$$

Usually, when the constriction factor is used, φ is set to 4.1 ($c_1 = c_2 = 2.05$), which gives the value of constriction factor K as 0.729. Carlisle and Dozier [58] showed that cognitive and social values of $c_1 = 2.8$ and $c_2 = 1.3$ also yield good results for their chosen set of problems.

The constriction approach is effectively equivalent to the inertia weight approach. Both approaches have the objective of balancing exploration and exploitation, and thereby improving convergence time and the quality of solution found. It should be noted that low values of ω and K result in exploitation with little exploration, while large values result in exploration with difficulties in refining solutions [51].

2.7 Time Series

A time series is a set of observations for the sequence of data x_t each one being recorded at a specific time. A discrete time series is one in which the set T_O of times at which observations are made is a discrete set, as is the case, for example, when observations are made at fixed time intervals. A continuous time series are obtained when observations are recorded continuously over some time interval, e.g., when $T_0 \in [0, 1]$

Loosely speaking, a time series $\{X_t, t = 0, \pm 1, \ldots\}$ is said to be stationary if it has statistical properties similar to those of the "time-shifted" series $\{X_t, t = 0, \pm 1, \ldots\}$, for each Restricting attention to those properties that depend only on the first-and second-order moments of $\{X_t\}$, we can make this idea precise with the following definitions. Let $\{X_t\}$ be a time series with $\left[EX_t^2\right] < \infty$. Then mean function of $\{X_t\}$ is $\mu_x(t) = E(X_t)$.

The covariance function of $\{X_t\}$ is:

$$Y_x(r, s) = Cov(X_r, X_s) = E[(X_r - \mu_x(r))(X_s - \mu_x(s))] \qquad (2.17)$$

for all integer r and s.

$\{X_t\}$ is (weakly) stationary if:

(i) $\mu_x(t)$ is independent of t.
(ii) $Y_x(t + h(t))$ is independent of t for each t.

Remark 1 Strict stationary time series $\{X_1, t = 0, \pm 1, \ldots\}$ is defined by the condition that $(X_1 \ldots X_n)$ and $(X_{(1+h)} \ldots X_{(n+h)})$ have some joint distributions for all integers h and $n > 0$. It is easy to check that if $\{X_t\}$ is also weakly stationary. Whenever we the use term stationary well shall mean weakly stationary.

Remark 2 In view of condition (ii), whenever we use term covariance function with reference to a stationary time series $\{X_t\}$ we shall mean the function Y_x of one variable, defined [40, 59–64] as:

$$Y_x(h) := Y_x(h, 0) = Y_x(t + h, t) \qquad (2.18)$$

The function $Y_x (.)$ will be referred to as the auto covariance function and $Y_x(h)$ as its value at log h [60].

Let $\{X_t\}$ be a stationary time series. The autocovariance function (ACVF) of $\{X_t\}$ at log h is:

$$Y_x(h) = Cov(X_{(1+h)}, X_t) \qquad (2.19)$$

The autocorrelation function (ACF) of $\{X_t\}$ at log h is

$$p_x \equiv \frac{Y_x(h)}{Y_x(0)} = Cor(X_{(1+h)}, X_t) \qquad (2.20)$$

2.8   Human Iris Biometrics

The biometrics systems can operate like a identification system or like a verification system. In the first case the objective of the system is identify the individual taking the biometric characteristic and comparing with the biometric characteristics existents in the database. In the first case, the objective of the system is identify the individual, this action is achieved with the obtaining the biometric characteristic of the individual and compares with the biometric characteristics in existence in the database.

In the second case the individual provide at system their identity, the system compare their biometric characteristic that is associated with the database to confirm that the individual really is he.

The first use of the iris was presented in Paris, where criminals were classified according to the color of their eyes following a proposal by the French ophthalmologist Bertillon [65]. Research in visual identification technology began in 1935. During that year an article appeared in the 'New York State Journal of Medicine', which suggested that "the pattern of arteries and veins of the retina could be used for unique identification of an individual" [66].

After researching and documenting the potential use of the iris as a tool to identify people, ophthalmologists Flom and Safir patented their idea in 1987 [67]; and later, in 1989, they patented algorithms developed with the mathematician Daugman. Thereafter, other authors developed similar approaches [66].

In 2001, Daugman also presented a new algorithm for the recognition of people using the biometric measurement of Iris [68].

In 2005, Roy proposes an iris recognition system for the identification of persons using support vector machine [69].

In 2006, Cho and Kim presented a new method to determine the winner in LVQ neural network [70].

In 2009, Sarhan used the discrete cosine transform for feature extraction and artificial neural network for recognition [71]; Abiyev and Altunkaya presented the iris recognition system using neural network [72].

The literature has well documented the uniqueness of visual identification. The iris is so unique that there are no two irises alike, even twins, in all humanity. The probability of two irises producing the same code is 1 in 1078, becoming known that the earth's population is estimated at approximately 1010 million [73], it is almost impossible to occur.

2.9   Historical Development

The backpropagation algorithm and its variations are the most useful basic training methods in the area of research of neural networks. However, these algorithms are too slow for practical applications.

When applying the basic backpropagation algorithm to practical problems, the training time can be very high. In the literature we can find that several methods have seen proposed to accelerate the convergence of the algorithm [3–5].

There exists many works about adjustment or managing of weights but only the most important and relevant for this research will be considered here.

Momentum method—Rumelhart, Hinton y Williams suggested adding in the increased weights expression a momentum term β, to filter the oscillations that can be formed a higher learning rate that lead to great change in the weights [7, 74].

Adaptive learning rate—focuses on improving the performance of the algorithm by allowing the learning rate changes during the training process (increase or decrease) [74].

Conjugate Gradient algorithm—A search of weight adjustment along conjugate directions. Versions of conjugate gradient algorithm differ in the way in which a constant βk is calculated.

- Fletcher-Reeves Update [75].
- Polak-Ribiere Update [75].
- Powell-Beale Restart [76, 77].
- Scaled Conjugate Gradient [78].

Kamarthi and Pittner [79], focused in obtaining a weight prediction of the network at a future epoch using extrapolation.

Ishibuchi et al. [80], proposed a fuzzy network where the weights are given as trapezoidal fuzzy numbers, denoted as four trapezoidal fuzzy numbers for the four parameters of trapezoidal membership functions.

Ishibuchi et al. [81], proposed a fuzzy neural network architecture with symmetrical fuzzy triangular numbers for the fuzzy weights and biases, denoted by the lower, middle and upper limit of the fuzzy triangular numbers.

Feuring [82], based on the work by Ishibuchi where triangular fuzzy weights are used, developed a learning algorithm in which the backpropagation algorithm is used to compute the new lower and upper limits media weights. The modal value of the new fuzzy weight is calculated as the average of the new computed limits.

Castro et al. [83], uses interval type 2 fuzzy neurons for the antecedents and interval of type 1 fuzzy neurons for the consequences of the rules. This approach handles the weights as numerical values to determine the input of the fuzzy neurons, as the scalar product of the weights for the input vector.

Gedeon [84], using a discrete selection (following the Hebbian paradigm) performs the weight adjustment. Monirul and Murase [85], as a strategy used the same weights in epochs where the output does not change. Meltser et al. [86], performed a weight adjustment of the network through the BFGS Quasi-Newton method

(Broyden-Fletcher-Goldfarn-Shanno). Barbouinis and Theocharis [87], performed the weights updating using the identification of recursive error prediction (RPE). Yeung et al. [88], used a new training objective function to adjust the weights for a network with radial basis functions.

Neville et al. [89], worked with sigma-pi networks, which are transformed for performing a second task of assignation for which they were initially trained. Casasent and Natarajan [90] used weights with complex values and the non linear square law. Yam and Chow [91] developed an algorithm to find the initial optimal weights of feed forward neural networks based on the Cauchy inequality and a linear algebraic method. Draghici [92], calculates a range of weights for a category of given problems and ensures that the network has the capacity to solve the given problems with integer weights in that range.

There are also recent works on type-2 fuzzy logic that have been developed in time series prediction, like that of Castro et al. [93], and other researchers [29, 94, 95].

In literature, the bio-inspired algorithms have proved that using them for optimization in different areas of research is possible find an optimal results. There are many works of different bio-inspired algorithms, but only works of the genetic algorithm and particle swarm optimization and the most important and relevant for this paper will be considered here: Valdez et al. [96], performed a hybridization of GA and PSO algorithm and integral the concept of fuzzy logic in it, and other researchers [97, 98].

References

1. Ni, H., Yiu, H.: Exchange rate prediction using hybrid neural networks and trading indicators. Neurocomputing **72**(13–15), 2815–2823 (2009)
2. Dow, J.: Indexes. http://www.djindexes.com. Accessed 5 Sept 2010
3. Cazorla, M., Escolano, F.: Two Bayesian methods for junction detection. IEEE Trans. Image Process. **12**(3), 317–327 (2003)
4. Martinez, G., Melin, P., Bravo, D., Gonzalez, F., Gonzalez, M.: Modular neural networks and fuzzy Sugeno integral for face and fingerprint recognition. Adv. Soft Comput. **34**, 603–618 (2006)
5. De Wilde, P.: The magnitude of the diagonal elements in neural networks. Neural Netw. **10**(3), 499–504 (1997)
6. Salazar, P.A., Melin, P., Castillo, O.: A new biometric recognition technique based on hand geometry and voice using neural networks and fuzzy logic. Soft Comput. Hybrid Intell. Syst. 171–186 (2008)
7. Phansalkar, V.V., Sastrq, P.S.: Analysis of the back-propagation algorithm with momentum. IEEE Trans. Neural Netw. **5**(3), 505–506 (1994)
8. Park, Y.N., Murray, T.J., Chen, C.: Predicting sun spots using a layered perceptron neural network. IEEE Trans. Neural Netw. **1**(2), 501–505 (1996)
9. Sharkey, A.: Combining Artificial Neural Nets: Ensemble and MODUlar Multi-Net Systems. Springer, London (1999)
10. Sharkey, A.: One combining Artificial of Neural Nets. Department of Computer Science University of Sheffield, U.K. (1996). Shimshoni, Y.: Intrator classification of seismic signal by

integrating ensemble of neural networks. IEEE Trans. Signal Process. **461**(5), 1194–1201 (1998)

11. Hansen, L.K., Salomon, P.: Ensemble methods for handwritten digit recognition. In: Proceedings of IEEE Workshop on Neural Networks for Signal Processing, Helsingoer, Denmark, pp. 333–342. IEEE Press, Piscataway, NJ (1992)

12. Goldberg, D.: Genetic Algorithms in Search, Optimization and Machine Learning. Addison Wesley (1989)

13. Gutta, S., Wechsler, H.: Face recognition using hybrid classifier systems. In: Proceedings of ICNN-96, Washington, DC, pp. 1017–1022. IEEE Computer Society Press, Los Alamitos, CA (1996)

14. Hansen, L.K., Salomon, P.: "Neural network ensembles. IEEE Trans. Pattern Anal. Mach. Intell. **12**(10), 993–1001 (1990)

15. Drucker, H., Schapire, R., Simard P.: Improving performance in neural networks using a boosting algorithm. In: Hanson, S.J., Cowan, J.D., Giles, C.L. (eds) Advances in Neural Information Processing Systems 5, Denver, CO, Morgan Kaufmann, San Mateo, CA, pp. 42–49 (1993)

16. Cherkauer, K.J.: Human expert level performance on a scientific image analysis task by a system using combined artificial neural networks. In: Chan, P., Stolfo, S., Wolpert, D. (eds) Proceedings of AAAI-96 Workshop on Integrating Multiple Learned Models for Improving and Scaling Machine Learning Algorithms, Portland, OR, pp. 15–21. AAAI Press, Menlo Park, CA (1996)

17. Cunningham, P., Carney, J., Jacob, S.: Stability problems with artificial neural networks and the ensemble solution. Artif. Intell. Med. **20**(3), 217–225 (2000)

18. Mencattini, A., Salmeri, M., Mertazzoni, B., Lojacono, R., Pasero, E., Moniaci, W.: Local meteorological forecasting by type-2 fuzzy systems time series prediction. In: CIMSA 2005— IEEE International Conference on Computational Intelligence for Measurement Systems and Applications Giardini Naxos, Italy, pp. 20–22 (2005)

19. Singh, S.R.: A simple method of forecasting based on fuzzy time series. Appl. Math. Comput. **186**(1), 330–339 (2007)

20. Hampshire, J., Waibel, A.: A novel objective function for improved phoneme recognition using time- delay neural networks. IEEE Trans. Neural Netw. **1**(2), 216–228 (1990)

21. Wei, L.-Y., Cheng C.-H.: A hybrid recurrent neural networks model based on synthesis features to forecast the Taiwan stock market. Int. J. Innov. Comput. Inf. Control **8**(8), 5559–5571

22. Maclin, R., Shavlik, J.W.: Combining the predictions of multiple classifiers: using competitive learning to initialize neural networks. In: Proceedings of IJCAI-95, Montreal, Canada, Morgan Kaufmann, San Mateo, CA, pp. 524–530 (1995)

23. Liu, F., Quek, C., See, G.. Ng: Neural network model for time series prediction by reinforcement learning. In: Proceedings of the International Joint Conference on the Neural Networks, Montreal, Canada (2005)

24. Soltani, S.: On the use of the wavelet decomposition for time series prediction. Neurocomputing **48**(1–4), 267–277 (2002)

25. Krogh, A., Vedelsby, J.: Neural network ensembles, cross validation, and active learning. In: Tesauro, G., Touretzky, D., Leen, T. (eds.) Advances in Neural Information Processing Systems 7, Denver, CO, pp. 231–238. MIT Press, Cambridge, MA (1995)

26. Opitz, D.W., Shavlik, J.W.: Generating accurate and diverse members of a neural network ensemble. In: Touretzky, D.S., Mozer, M.C., Hasselmo, M.E. (eds.) Advances in Neural Information Processing Systems 8, Denver, CO, pp. 535–541. MIT Press, Cambridge, MA (1996)

27. Yao, X., Liu, F.: Evolving neural network ensembles by minimization of mutual information. Int. J. Hybrid Intell. Syst. **1**, 12–21 (2004)

28. Xue, J., Xu, Z., Watada, J.: Building an integrated hybrid model for short-term and mid-term load forecasting with genetic optimization. Int. J. Innov. Comput. Inf. Control **8**(10), 7381–7391 (2012)

29. Karnik, N., Mendel, M.: Applications of type-2 fuzzy logic systems to forecasting of time-series. Inf. Sci. **120**(1–4), 89–111 (1999)
30. Karnik, N., Mendel, M.: "Introduction to type-2 fuzzy logic systems. IEEE Trans. Signal Process. **2**, 915–920 (1998)
31. Karnik, N., Mendel, J.M.: Operations on type-2 set. Fuzzy Set Syst. **122**, 327–348 (2001)
32. Tong, R.M., Nguyen, H.T.: Fuzzy sets and applications: selected papers by L.A. Zadeh. In: Yager, R.R., Ovchinnikov, S. (eds) Wiley, New York (1987)
33. Zadeh, L.A.: Fuzzy sets, information and control. **8**, 338–353 (1965)
34. Fazel Zarandi, M.H., Rezae, B., Turksen, I.B., Neshat, E.: A type-2 fuzzy rule-based expert system model for stock price analysis. Expert Syst. Appl. **36**, 139–154 (2009)
35. Jang, J.S.R., Sun, C.T., Mizutani, E.: Neuro-Fuzzy and Soft Computing: A Computational Approach to Learning and Machine Intelligence, p. 614. Prentice Hall, (1997)
36. Jilani, T.A., Burney, S.: A refined fuzzy time series model for stock market forecasting. Phys. A Stat. Mech. Appl. **387**, 2857–2862 (2008)
37. Mendel, J.: Uncertain Rule-Based Fuzzy Logic Systems. Introduction an New Directions, pp. 213–231. Prentice-Hall Inc. (2001)
38. Eberhart, R., Shi, Y., Kennedy, J.: "Swam Intelligence", San Mateo. Morgan Kaufmann, California (2001)
39. Andreas, A., Sheng, W.: Introduction Optimization, Practical Optimization Algorithms and Engineering Applications, pp. 1–4. Springer (2007)
40. Antoniou, A., Sheng, W.: Practical optimization algorithms and engineering applications "introduction optimization". In: Antoniou, A., Sheng, W. (eds), pp. 1–4. Springer (2007)
41. Yen, J., Langari, R.: Fuzzy Logic: Intelligence, Control and Information. Prentice Hall (1999)
42. Dorigo, M., Stutzle, T.: Ant colony optimization "ant colony optimization theory". A Bradford Book The Milt Press London England, Massachusetts Institute of Technology, pp. 121–151 (2004)
43. Valle, Y., Venayagamoorthy, G.K., Mohagheghi, S., Hernandez, J.-C., Harley, R.G.: Particle swarm optimization: basic concepts, variants and applications in power systems. In: IEEE Transactions on Evolutionary Computation, pp. 171–195 (2008)
44. Holland, J.: Adaptation in natural and artificial systems. University of Michigan Press (1975)
45. Eberhart, R, Shi, Y.: Comparing inertia weights and constriction factors in particle swarm optimization. In: Proceedings of the IEEE Congress on Evolutionary Computation, vol. 1, pp. 84–88 (2000)
46. Eberhart, R., Shi, Y.: Comparison between genetic algorithms and particle swarm optimization. In: Proceedings of the Seventh Annual Conference on Evolutionary Programming, pp. 611–616 (1998)
47. Mao, J.: A case study on bagging, boosting and basic ensembles of neural networks for OCR. In: Proceedings of IJCNN-98, Anchorage, AK, vol. 3, pp. 1828–1833. IEEE Computer Society Press, Los Alamitos, CA (1998)
48. Reeves, C., Row, J.: Genetic algorithms: principles and perspectives, "optimization", pp. 4–8. Kluwer Academic Publishers, New York (2002)
49. Whitley, L.D.: Foundations of Genetic Algorithms, vol. 2, p. 332. Morgan Kaufman Publishers (1993)
50. Eberhart, R., Kennedy, J.: A new optimizer using particle swarm theory. In: Proceedings of 6th International Symposium on Micro Machine and Human Science (MHS), pp. 39–43 (1995)
51. Engelbrech. P.: Fundamentals of Computational of Swarm Intelligence: Basic Particle Swarm Optimization, pp. 93–129. Wiley (2005)
52. Escalante, H.J., Montes, M., Sucar, L.E.: Particle swarm model selection. J. Mach. Learn. Res. **10**, 405–440 (2009)
53. Kennedy, J., Eberhart, R.: Particle swam optimization. In: Proceedings of IEEE International Conference on Neural Network (ICNN), vol. 4, pp. 1942–1948 (1995)
54. Kim, D., Kim, C.: Forecasting time series with genetic fuzzy predictor ensemble. IEEE Trans. Fuzzy Syst. **5**(4) (1997)

55. Shi, Y., Eberhart, R.C.: A modified particle swarm optimizer. In: Proceedings of the IEEE Congress of Evolutionary Computation, pp. 69–73 (1998)
56. Clerc, M.: The swarm and the queen: towards a deterministic and adaptive particle swarm optimization. In: Proceedings of the IEEE Congress on Evolutionary Computation, vol. 3, pp. 1951–1957 (1999)
57. Clerc, M., Kennedy, J.: The particle swarm-explosion, stability, and convergence in a multimodal complex space. IEEE Trans. Evol. Comput. **6**, 58–73 (2002)
58. Carlisle, Dozier, G.: An off-the-shelf PSO. In: Proceedings of the Workshop on Particle Swarm Optimization, Indianapolis, USA, pp. 1–6 (2001)
59. Brockwell, P.D., Davis, R.A.: Introduction to Time Series and Forecasting, pp. 1–219. Springer, New York (2002)
60. Cowpertwait, P., Metcalfe, A.: Time Series. Introductory Time Series with R, pp. 2–5. Springer, Dordrecht (2009)
61. Davey, N., Hunt, S., Frank, R.: Time Series Prediction and Neural Networks. University of Hertfordshire, Hatfield, UK (1999)
62. Plummer, E.A.: Time series forecasting with feed-forward neural networks: guidelines and limitations. University of Wyoming (2000)
63. Wei, W.W.S.: Time Series Analysis: Univariate and Multivariate Methods, vol. 1, pp 40–100. Addison-Wesley (1994)
64. Zhang, J., Man, K.F.: Time series prediction using recurrent neural network in multi-dimension embedding phase space. IEEE Int. Conf. Syst. Man Cybernet. **2**, 11–14 (1998)
65. Tisse, C., Torres, L.M., Robert M.: Person identification technique using human iris recognition. Universidad de Montepellier (2000)
66. López, J.,: Estado del Arte: Reconocimiento Automático del Iris Humano", Politécnico Colombiano, y Javier González Patiño, Universidad Nacional de Colombia, Scientia et Technica Año XI, No 29, pp. 77–81 (2005)
67. Khaw, P.: Iris recognition technology for improved authentication, pp. 1–17. Sala de Lectura de Seguridad de la Información, SANS Institute (2002)
68. Daugman, J.: Statistical richness of visual phase information: update on recognizing persons by iris patterns. Int. J. Comput. Vis. **45**(1), 25–38 (2001)
69. Roy, K., Bhattacharya, P.: Iris recognition with support vector machines. Advances in biometrics. Lect. Notes Comput. Sci. **3822**, 486–492 (2005)
70. Cho, S., Kim, J.: Iris recognition using LVQ neural network. Adv. Neural Netw. Lect. Notes Comput. Sci. **3972**, 26–33 (2006)
71. Sarhan, A.: Iris recognition using discrete cosine transform and artificial neural networks. J. Comput. Sci. **5**, 369–373 (2009)
72. Abiyev, R., Altunkaya, K.: Neural network based biometric personal identification with fast iris segmentation. Int. J. Control Autom. Syst. **7**(1), 17–23 (2009)
73. Sánchez, O., y González, J.: Control de Acceso basado en Reconocimiento de Iris. Corporación Universitaria Tecnológica de Bolívar, Facultad de Ingeniería Eléctrica, Electrónica y Mecatrónica, Cartagena de Indias, Monografía, Noviembre 2003, pp. 1–137
74. Hagan, M.T., Demuth, H.B., Beale M.H.: Neural Network Design, pages 736. PWS Publishing (1996)
75. Fletcher, R., Reeves, C.M.: Function minimization by conjugate gradients. Comput. J. **7**, 149–154 (1964)
76. Powell, M.J.D.: Restart procedures for the conjugate gradient method. Math. Program. **12**, 241–254 (1977)
77. Beale, E.M.L.: "A Derivation of Conjugate Gradients. In: Lootsma, F.A. (ed.) Numerical methods for nonlinear optimization, pp. 39–43. Academic Press, London (1972)
78. Moller, M.F.: A scaled conjugate gradient algorithm for fast supervised learning. Neural Netw. **6**, 525–533 (1993)
79. Kamarthi, S., Pittner, S.: Accelerating neural network training using weight extrapolations. Neural Netw. **12**(9), 1285–1299 (1999)

80. Ishibuchi, H., Morioka, K., Tanaka, H.: A fuzzy neural network with trapezoid fuzzy weights. In: Fuzzy Systems, vol. 1, IEEE World Congress on Computational Intelligence, pp. 228–233 (1994)
81. Ishibuchi, H., Tanaka, H., Okada, H.: Fuzzy neural networks with fuzzy weights and fuzzy biases. In: IEEE International Conference on Neural Networks, vol. 3, pp. 160–165 (1993)
82. Feuring, T.: Learning in fuzzy neural networks, neural networks. In: IEEE International Conference on, vol. 2, pp. 1061–1066 (1996)
83. Castro, J., Castillo, O., Melin, P., Rodríguez-Díaz, A.: A hybrid learning algorithm for a class of interval type-2 fuzzy neural networks. Inf. Sci. **179**(13), 2175–2193 (2009)
84. Gedeon, T.: Additive neural networks and periodic patterns. Neural Netw. **12**(4–5), 617–626 (1999)
85. Monirul Islam, Md., Murase, K.: A new algorithm to design compact two-hidden-layer artificial neural networks. Neural Netw. **14**(9), 1265–1278 (2001)
86. Meltser, M., Shoham, M., Manevitz, L.: Approximating functions by neural networks: a constructive solution in the uniform norm. Neural Netw. **9**(6), 965–978 (1996)
87. Barbounis, T.G., Theocharis, J.B.: Locally recurrent neural networks for wind speed prediction using spatial correlation. Inf. Sci. **177**(24), 5775–5797 (2007)
88. Yeung, D., Chan, P., Ng, W.: Radial basis function network learning using localized generalization error bound. Inf. Sci. **179**(19), 3199–3217 (2009)
89. Neville, R.S., Eldridge, S.: Transformations of Sigma–Pi nets: obtaining reflected functions by reflecting weight matrices. Neural Netw. **15**(3), 375–393 (2002)
90. Casasent, D., Natarajan, S.: A classifier neural net with complex-valued weights and square-law nonlinearities. Neural Netw. **8**(6), 989–998 (1995)
91. Yam, J., Chow, T.: A weight initialization method for improving training speed in feedforward neural network. Neurocomputing **30**(1–4), 219–232 (2000)
92. Draghici, S.: On the capabilities of neural networks using limited precision weights. Neural Netw. **15**(3), 395–414 (2002)
93. Castro, J., Castillo, O., Melin, P., Mendoza, O., Rodríguez-Díaz, A.: An interval type-2 fuzzy neural network for chaotic time series prediction with cross-validation and Akaike test. Soft Comput. Intell. Control Mobile Robot. 269–285 (2011)
94. Abiyev, R.: A type-2 fuzzy wavelet neural network for time series prediction. Lect. Notes Comput. Sci. **6098**, 518–527 (2010)
95. Pulido, M., Melin, P., Castillo, O.: Genetic optimization of ensemble neural networks for complex time series prediction. IJCNN 202–206 (2011)
96. Valdez, F., Melin, P., Castillo, O.: Evolutionary method combining particle swarm optimization and genetic algorithms using fuzzy logic for decision making. In: Proceedings of the IEEE International Conference on Fuzzy Systems, pp. 2114–2119 (2009)
97. Sanchez, D., Melin, P.: Modular neural network with fuzzy integration and its optimization using genetic algorithms for human recognition based on iris, ear and voice biometrics. Soft Comput. Recogn. Based Biometrics 85–102 (2010)
98. Valdez, F., Melin, P., Parra, H.: Parallel genetic algorithms for optimization of modular neural networks in pattern recognition. IJCNN 314–319 (2011)

Chapter 3
Problem Statement and Development

The proposed approach in this book has the goal of generalizing the backpropagation algorithm using type-1 fuzzy sets and type-2 fuzzy sets to allow the neural network to handle data with uncertainty. In the type-2 fuzzy sets, it will be necessary vary the footprint of uncertainty (FOU) of the membership functions using an optimization method to make it automatically or vary it manually for the corresponding applications [1–4].

The process of obtaining the weights in the connections for each neuron is performed differently to the traditional adjustment of weights with the backpropagation algorithm (Fig. 3.1).

The proposed method works with type-1 and type-2 fuzzy weights, considering the possible modification in the way we work internally in the neuron and the adjustment of the weights given in this way (Figs. 3.2 and 3.3) [5, 6].

We will consider the option of adapting the current methods of adjusting weights that allow convergence to the correct weights for the problem. To overcome difficulties with the current methods, we develop a method for adjusting weights to achieve the desired result, searching for the optimal way to work with type-2 fuzzy weights [7, 8].

In the literature can be found that it has very difficult and exhaustive to find optimal values for a problem manually, so this raises the use of bio-inspired methods for obtaining the optimal type 2 fuzzy weights for the neural network [9].

Fuzzy weights are denoted as follows.

$$\widetilde{w_{kj}} = \left[\underline{w_{kj}}, \overline{w_{kj}}\right], \widetilde{w_{jl}} = \left[\underline{w_{jl}}, \overline{w_{jl}}\right], \ldots, \widetilde{w_{jn}} = \left[\underline{w_{jn}}, \overline{w_{jn}}\right]. \tag{3.1}$$

In Fig. 3.4 an example of neural network architecture with type 2 fuzzy weights is shown:

We used the sigmoid activation function for the hidden neurons and the linear activation function for the output neurons, and we utilized this activation functions because these functions have obtained good results in similar approaches.

© The Author(s) 2016
F. Gaxiola et al., *New Backpropagation Algorithm with Type-2 Fuzzy Weights for Neural Networks*, SpringerBriefs in Computational Intelligence,
DOI 10.1007/978-3-319-34087-6_3

Fig. 3.1 Scheme current
management of numerical
weights (type 0) for input of
each neuron

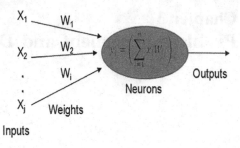

Fig. 3.2 Scheme of the
proposed management of
type-1 fuzzy weights for the
inputs of each neuron

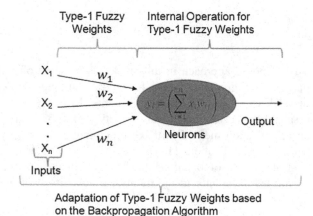

Fig. 3.3 Scheme of the
proposed management of
type-2 fuzzy weights for the
inputs of each neuron

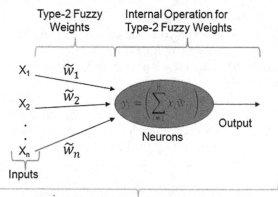

Fig. 3.4 Example of neural network architecture with type 2 fuzzy weights

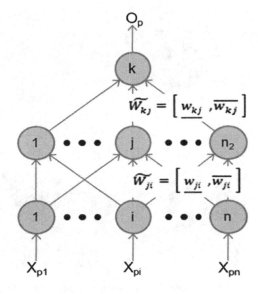

3.1 Historical Data

The data used in this book are the Mackey-Glass for $\tau = 17$ for time series prediction and the CASIA iris database for the recognition.

3.1.1 Iris Database

We used a database of human Iris from the Institute of Automation of the Chinese Academy of Sciences (CASIA) (see Fig. 3.5). It consists of 9 images per person, for a total of 10 individuals, giving a total of 90 images. The image dimensions are 320×280, JPEG format [10–12].

3.1.2 Mackey-Glass Time Series

Where it is assumed x (0) = 1.2, $\tau = 17$, x (t) = 0 for t < 0. Figure 3.6 shows a plot of the time series for these parameter values.

This time series is chaotic, and there is no clearly defined period. The series does not converge or diverge, and the trajectory is extremely sensitive to the initial conditions. The time series is measured in number of points, and we apply the fourth order Runge-Kutta method to find the numerical solution of the equation [13–15].

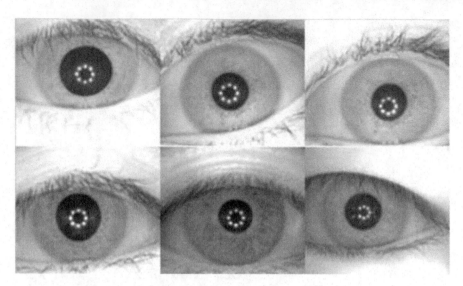

Fig. 3.5 Examples of the human iris images from the CASIA database

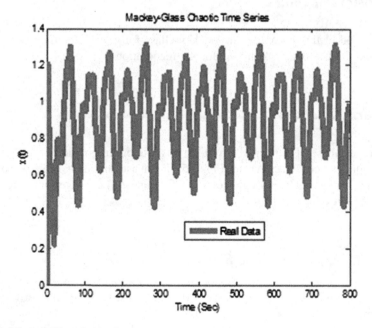

Fig. 3.6 Mackey-Glass time series

3.1.3 Dow-Jones Time Series

Dow Jones & Company is an American company that publishes financial reporting. It was founded in 1882 by three reporters: Charles Henry Dow, Edward David Jones, and Charles Bergstresser Milford. In that same year they began to publish a financial newsletter of only two sheets called "The Customer's Afternoon Letter" which would be the precursor of the famous financial newspaper The Wall Street Journal first released on July 8, 1889. The newsletter publicly showing the stock prices and the financial accounts of enterprises, the only information that had the people close to the company until then.

To better represent the movements of the stock market at the time, the Dow Jones & Company conducted a meter barometer of economic activity by creating twelve companies with the Dow Jones stock index. Likewise New York Times and Washington Post newspapers, the company is open to the bag but is controlled by private. So far, the company is controlled by the Bancroft family, which controls 64 % of the shares entitled to vote [16].

Data of the Dow Jones time series: The data consist of 800 points that correspond to a period from 09/11/2005 to 15/01/2009 (as shown in Fig. 3.7). In this case 50 % of the data are used for the ensemble neural network trainings and 50 % to test the network [17].

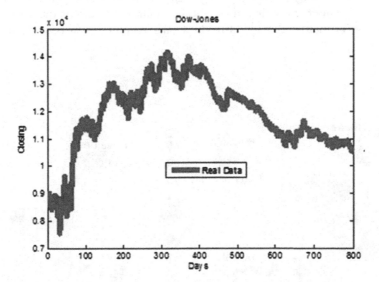

Fig. 3.7 Dow-Jones Time Series

3.2 Neural Network with Type-2 Fuzzy Weights Adjustment for Pattern Recognition of the Human Iris Bio-metrics

This work focuses primarily on developed of the method of the neural network with type-2 fuzzy weights and the identification of individuals. This problem is developed in many research in this area, considering various measures to achieve it with biometrics (fingerprint, voice, palm of hand, signature) and various methods of identification (with particular emphasis on neural networks).

The work was focused on the recognition of persons using a modular neural network with type-2 fuzzy weights, and image preprocessing methods to obtain the interest region of the iris.

The images of the human iris introduced to the two neural networks were pre-processed as follows:

- Obtain the coordinates and radius of the iris and pupil.
- Making the cut in the Iris.
- Resize the cut of the Iris to 21-21 pixels.
- Convert images from vector to matrix
- Normalize the images.

The proposed architecture with two neural networks with type-2 fuzzy weights consist of 120 neurons in the hidden layer and 10 neurons in the output layer, the inputs are the preprocessed iris images with a total of 10 persons (60 for training −60 for test in total) (see Fig. 3.8). The inputs vary in ±5 % between the two networks.

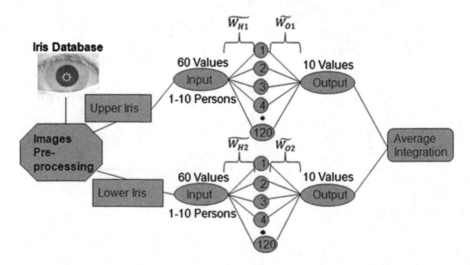

Fig. 3.8 Proposed neural network architecture with type-2 fuzzy weights for iris biometric

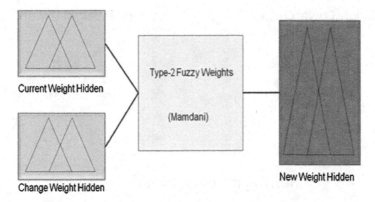

Fig. 3.9 Structure for the four type-2 fuzzy integration system for iris biometric

We worked with two neural networks those managing type-2 fuzzy weights in each hidden layer and output layer. In each hidden layer and output layer worked with a type-2 fuzzy inference system to obtain the new weights in each epoch of the network.

The two neural networks used the learning method that updates weight and bias values according to the resilient backpropagation algorithm. The updates weights are adapted for manage type-2 fuzzy weights.

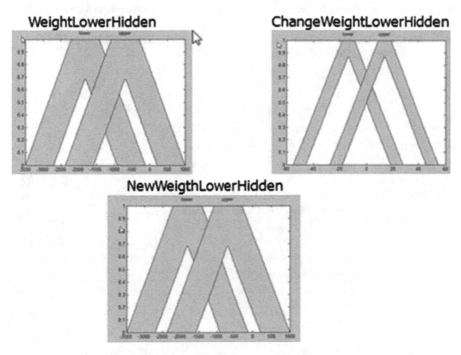

Fig. 3.10 Inputs and Output for the type-2 fuzzy inference system for the hidden layer in the first network for iris biometric

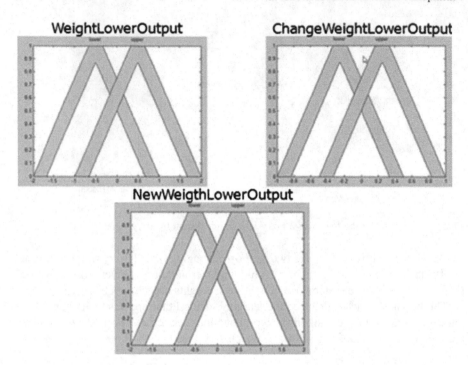

Fig. 3.11 Inputs and Output for the type-2 fuzzy inference system for the output for iris biometric

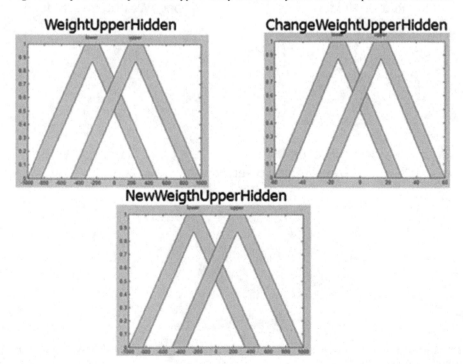

Fig. 3.12 Inputs and Output for the type-2 fuzzy inference system for the hidden layer in the second network for iris biometric

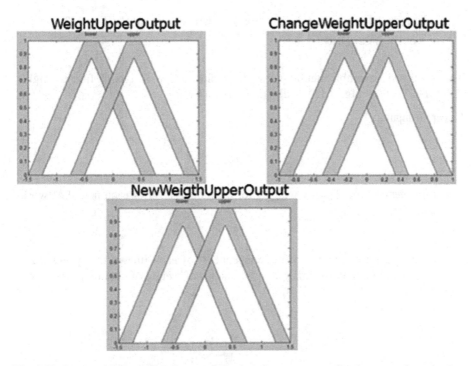

Fig. 3.13 Inputs and Output for the type-2 fuzzy inference system for the output layer in the second network for iris biometric

We used four type-2 fuzzy inference systems to obtain the new weights, one for the hidden layer in the one network and the second network, and one for the output layer in the first network and the second network.

The four type-2 fuzzy inference system consists of two inputs (actual weight and change of weight) and one output (new weight) (see Fig. 3.9).

The two inputs have two triangular membership functions, and the output have two triangular membership functions: for the hidden layer in the first network (see Fig. 3.10), for the output layer in the first network (see Fig. 3.11), for the hidden layer in the second network (see Fig. 3.12), and for the output layer in the second network (see Fig. 3.13); and work with six rules (see Fig. 3.14).

The integration of the two networks is realized with average integration.

1. If (WeigthLowerInput is lower) and (DelthaLowerInput is lower) then (NewWeigthLowerInput is lower) (1)
2. If (WeigthLowerInput is lower) and (DelthaLowerInput is upper) then (NewWeigthLowerInput is upper) (1)
3. If (WeigthLowerInput is upper) and (DelthaLowerInput is lower) then (NewWeigthLowerInput is lower) (1)
4. If (WeigthLowerInput is upper) and (DelthaLowerInput is upper) then (NewWeigthLowerInput is upper) (1)
5. If (WeigthLowerInput is lower) then (NewWeigthLowerInput is lower) (1)
6. If (WeigthLowerInput is upper) then (NewWeigthLowerInput is lower) (1)

Fig. 3.14 Rules for the four type-2 fuzzy inference system for iris biometric

3.3   Ensemble Neural Network Architecture with Type-2 Fuzzy Weights

The proposed ensemble neural network architecture with type-2 fuzzy weights (ENNT2FW) (see Fig. 3.15) is described as follows:

Layer 0: Inputs.

$$x = [x_1, x_2, \ldots, x_n]. \qquad (3.2)$$

Layer 1: Interval type-2 fuzzy weights for the hidden layer of each neural network.

$$\tilde{w} = [\underline{w}, \bar{w}]. \qquad (3.3)$$

where $[\underline{w}, \bar{w}]$ are the weights of the consequents of each rule of the type-2 fuzzy system with inputs (current fuzzy weight, change of weight) and output (new fuzzy weight).

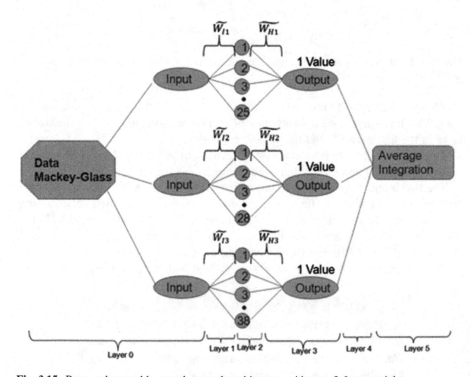

Fig. 3.15 Proposed ensemble neural network architecture with type-2 fuzzy weights

Layer 2: Hidden neuron with interval type-2 fuzzy weights.

$$\text{Net} = \sum_{i=1}^{n} x_i \widetilde{w_i}.$$ (3.4)

Layer 3: Output neuron with interval type-2 fuzzy weights.

$$\text{Out} = \sum_{i=1}^{n} y_i \widetilde{w_i}.$$ (3.5)

Layer 4: Obtain a single output of each one of the three neural networks.
Layer 5: Obtain a final output with the average integration.

We performed experiments in time-series prediction, specifically for the Mackey-Glass time series ($\tau = 17$).

We considered with three neural networks in the ensemble: the first network with 25 neurons in the hidden layer and 1 neuron in the output layer; the second network with 28 neurons in the hidden layer and 1 neuron in the output layer; and the third network with 38 neurons in the hidden layer and 1 neuron in the output layer. This ensemble neural network handles type-2 fuzzy weights in each one of its hidden layers and output layer. In each hidden layer and output of each network we are working with a type-2 fuzzy inference system getting new weights in each epoch of the network [18–21].

The combination of responses of the ensemble neural network is performed by average integration.

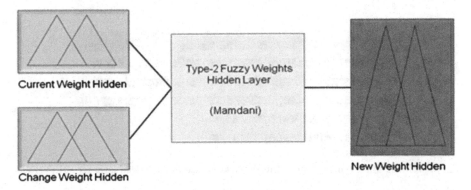

Fig. 3.16 Structure of the used type-2 fuzzy inference system in the hidden layer for ENNT2FW

We used 2 similar type-2 fuzzy systems in each neural network with 2 inputs and 1 output for the hidden layer (See Figs. 3.16 and 3.17) and 6 rules (see Fig. 3.18); and output layer (See Figs. 3.19 and 3.20) and 6 rules (see Fig. 3.21):

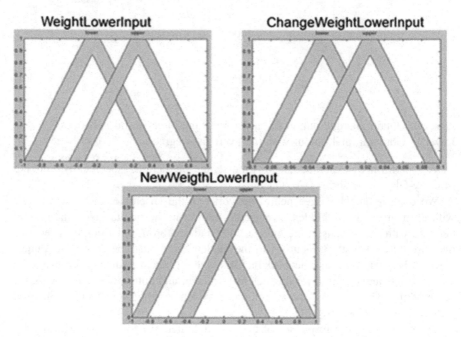

Fig. 3.17 Inputs and outputs of the type-2 fuzzy inference system for the hidden layer for ENNT2FW

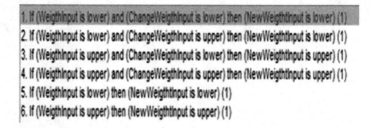

Fig. 3.18 Rules of the type-2 fuzzy inference systems used in the hidden layer for ENNT2FW

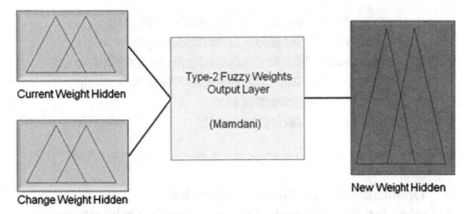

Fig. 3.19 Structure of the used type-2 fuzzy inference system in the output layer for ENNT2FW

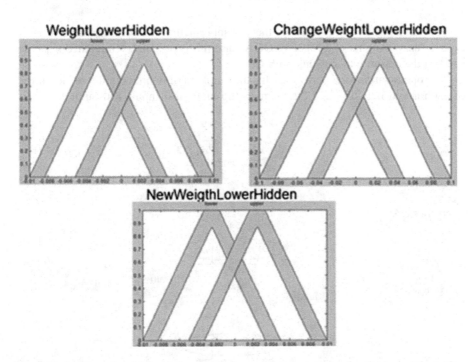

Fig. 3.20 Inputs and outputs of the type-2 fuzzy inference system for the output layer for ENNT2FW

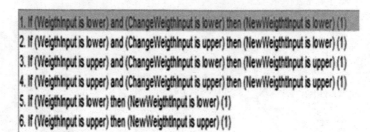

1. If (WeigthInput is lower) and (ChangeWeigthInput is lower) then (NewWeigthInput is lower) (1)

2. If (WeigthInput is lower) and (ChangeWeigthInput is upper) then (NewWeigthInput is lower) (1)

3. If (WeigthInput is upper) and (ChangeWeigthInput is lower) then (NewWeigthInput is upper) (1)

4. If (WeigthInput is upper) and (ChangeWeigthInput is upper) then (NewWeigthInput is upper) (1)

5. If (WeigthInput is lower) then (NewWeigthInput is lower) (1)

6. If (WeigthInput is upper) then (NewWeigthInput is upper) (1)

Fig. 3.21 Rules of the type-2 fuzzy inference systems used in the output layer for ENNT2FW

3.4 Optimization with Genetic Algorithm (GA) for the Ensemble Neural Network Architecture with Type-2 Fuzzy Weights

We performed the optimization of the ensemble neural network architecture with type-2 fuzzy weights (ENNT2FW-OGA) and the type-2 fuzzy inference systems used in the before experiments with GA.

The optimization was performed for the numbers of neurons in the hidden layer of each neural network, and the membership functions (inputs and outputs) of the

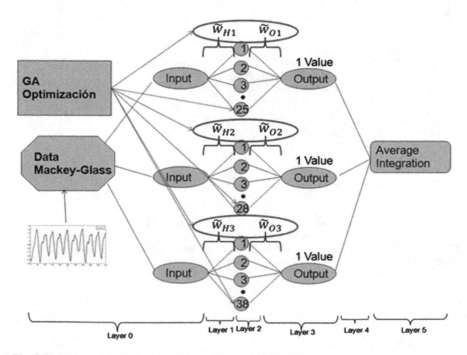

Fig. 3.22 Proposed optimization of the architecture ENNT2FW

Table 3.1 Parameters of the genetic algorithm used for optimization the ensemble neural network	Individuals	40
	Gens	81 (binary)
	Generations	20
	Assign fitness	Ranking
	Selection	Stochastic universal sampling
	Crossover	Single-point
	Mutation	0.0086

two type-2 fuzzy inference system used for obtaining the weights of the hidden layer and output layer, in Fig. 3.22 is described.

The parameters for the Genetic algorithm used to optimize the ensemble neural network are described in Table 3.1.

The ensemble neural network architecture obtained after the optimization is as follow: the first network with 30 neurons in the hidden layer and 1 neuron in the output layer; the second network with 29 neurons in the hidden layer and 1 neuron in the output layer; and the third network with 26 neurons in the hidden layer and 1 neuron in the output layer.

The membership functions for the two type-2 fuzzy inference system (inputs: current weight and change of weight, and output: new weight) optimized are show in the Fig. 3.23 for the hidden layer and Fig. 3.24 for the output layer.

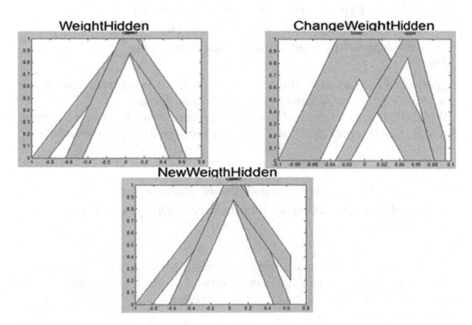

Fig. 3.23 Inputs and outputs optimized for the type-2 fuzzy inference system for the hidden layer for ENNT2FW

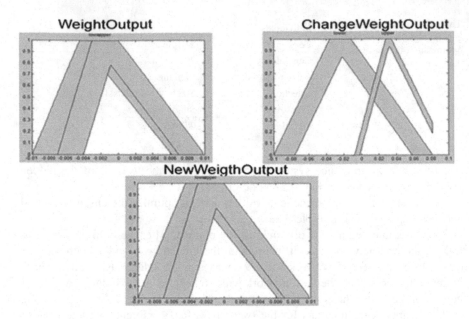

Fig. 3.24 Inputs and outputs optimized for the type-2 fuzzy inference system for the output layer for ENNT2FW

3.5   Interval Type-2 Fuzzy Weight Adjustment for Backpropagation Neural Networks with Application in Time Series Prediction

We performed a comparison of the monolithic neural network against the neural network with type-1 fuzzy weights and type-2 fuzzy weights.

We describe the architecture of the monolithic neural network, the neural network with type-1 fuzzy weights and the neural network with type-2 fuzzy weights as follows.

3.5.1   The Monolithic Neural Network Architecture and Neural Network Architecture with Type-1 Fuzzy Weights

The proposed monolithic neural network architecture (see Fig. 3.25) is described as follows: we used a monolithic neural network with the test data of the Mackey-Glass time series for the input layer, 30 neurons in the hidden layer and 1 neuron in the output layer.

The proposed neural network architecture with type-1 fuzzy weights (see Fig. 3.26) is described as follows:

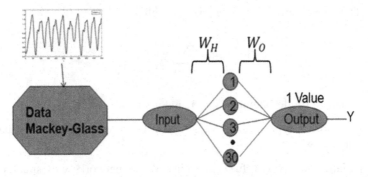

Fig. 3.25 Monolithic neural network architecture for Mackey-Glass

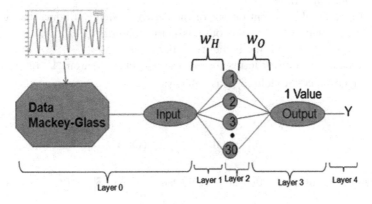

Fig. 3.26 Neural network architecture with type-1 fuzzy weights for Mackey-Glass

Layer 0: Inputs.

$$x = [x_1, x_2, \ldots, x_n] \tag{3.6}$$

Layer 1: Type-1 fuzzy weights for the hidden layer.

$$w = \frac{\sum_{i=1}^{n} (f_i w_i)}{\sum_{i=1}^{n} (f_i)} \tag{3.7}$$

Layer 2: Hidden neuron with type-1 fuzzy weights.

$$Net = \sum_{i=1}^{n} x_i w_i \tag{3.8}$$

Layer 3: Output neuron with type-1 fuzzy weights.

$$\text{Out} = \sum_{i=1}^{n} y_i w_i \tag{3.9}$$

Layer 4: Obtain a single output of the neural network.

The adaptation of type-1 fuzzy weights will be based on the backpropagation algorithm, as follows:

Step 1: Initialize the type-1 fuzzy weights of the network with small random values.

Step 2: Present an input pattern and specify the desired output that the network should generate.

Step 3: Calculate the current output of the network. First, introduce the network inputs and calculate the outputs corresponding to each layer until the output layer, and this is the network output.

Step 4: Calculate the error terms for all neurons. For a neuron "k" of the output layer, calculate delta δ_{pk}^{O}, as follows:

$$\delta_{pk}^{O} = \left(d_{pk} - y_{pk}\right) f_k^{O'}(\text{Out}) \tag{3.10}$$

For a neuron "j" of the hidden layer, calculate the delta $\delta_{pj}^{h} = f_j^{h'}$, as follows:

$$\delta_{pj}^{h} = f_j^{h'}(\text{Net}) \sum_k \delta_{pk}^{O} w_{kj} \tag{3.11}$$

Step 5: For updating the type-1 fuzzy weights using a recursive algorithm, starting from the output neurons and working back up until the input layer, adjusting the type-1 fuzzy weights as follows:

The update uses two type-1 fuzzy inference systems with 2 inputs (Current type-1 fuzzy weight w_{kj} and Change of type-1 fuzzy weight $\Delta w_{kj}(t+1)$ and 1 output (Resulting type-1 fuzzy weights $w_{kj}(t+1)$.

The change of type-1 fuzzy weights is performed with the following equations: For the neurons of the output layers:

$$\Delta w_{kj}(t+1) = \delta_{pk}^{O} y_{pj} \tag{3.12}$$

For the neurons of the hidden layers:

$$\Delta w_{ji}(t+1) = \delta_{pj}^h x_{pi} \tag{3.13}$$

Step 6: The process is repeated until the error terms are sufficiently small for each of the learned patterns.

$$E_p = \frac{1}{2}\sum_{k=1}^{M}\delta_{pk}^2. \tag{3.14}$$

We used 2 similar type-1 fuzzy systems to obtain the type-1 fuzzy weights in the hidden and output layer for the neural network.

The first type-1 fuzzy system consists of two inputs: the current weight in the actual epoch and the change of the weight for the next epoch, and one output: the new weight for the next epoch (see Fig. 3.27).

The input of the current weight has two triangular membership functions with range of −1 to 1. The input of change of the weight has two triangular membership functions with range of −0.003 to 0.003. The output of the new weight has two triangular membership functions with range of −1 to 1 (see Fig. 3.28).

We used six rules for the type-1 fuzzy inference system of the hidden layer, corresponding to the four combinations of the two triangular membership functions and we added two rules for when the change of weight is null (see Fig. 3.29).

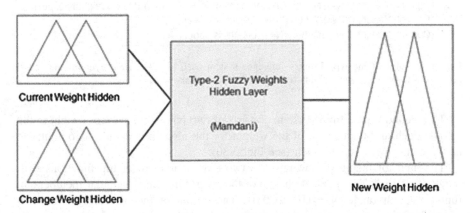

Fig. 3.27 Type-1 fuzzy inference system used in the hidden layer for the neural network with type-1 fuzzy weights (triangular MF) for Mackey-Glass

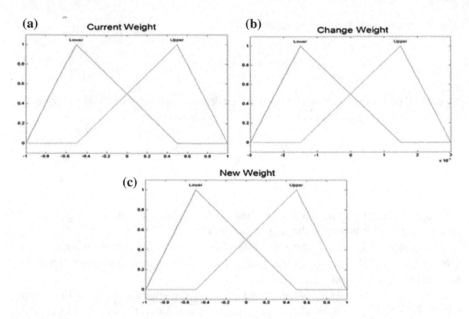

Fig. 3.28 Inputs (**a** and **b**) and output (**c**) of the type-1 fuzzy inference system used in the hidden layer for the neural network with type-1 fuzzy weights (triangular MF) for Mackey-Glass

---

1. (Current_Weight is lower) and (Change_Weight is lower) then (New_Weight is lower)
2. (Current_Weight is lower) and (Change_Weight is upper) then (New_Weight is lower)
3. (Current_Weight is upper) and (Change_Weight is lower) then (New_Weight is upper)
4. (Current_Weight is upper) and (Change_Weight is upper) then (New_Weight is upper)
5. (Current_Weight is lower) then (New_Weight is lower)
6. (Current_Weight is upper) then (New_Weight is upper)

Fig. 3.29 Rules of the type-1 fuzzy inference system used in the hidden layer for the neural network with type-1 fuzzy weights (triangular MF) for Mackey-Glass

The second type-1 fuzzy system consists of two inputs: the current weight in the actual epoch and the change of the weight for the next epoch, and one output: the new weight for the next epoch (see Fig. 3.30).

The input of the current weight has two triangular membership functions with range of −1 to 1. The input of change of the weight has two triangular membership functions with range of −0.01 to 0.01. The output of the new weight has two triangular membership functions with range of −1 to 1 (see Fig. 3.31).

We used the same six rules before presented for the type-1 fuzzy inference system for the output layer (see Fig. 3.29).

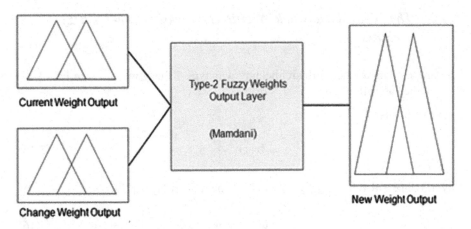

Fig. 3.30 Type-1 fuzzy inference system used in the output layer for the neural network with type-1 fuzzy weights (triangular MF) for Mackey-Glass

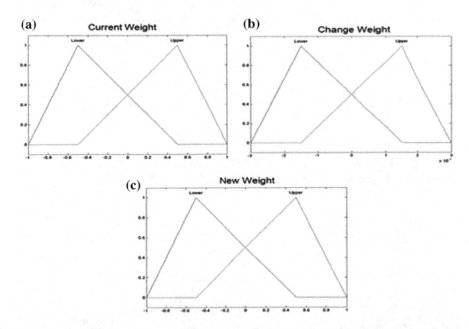

Fig. 3.31 Inputs (**a** and **b**) and output (**c**) of the type-1 fuzzy inference system used in the output layer for the neural network with type-1 fuzzy weights (triangular MF) for Mackey-Glass

3.5.2 The Neural Network Architecture with Type-2 Fuzzy Weights

The proposed neural network architecture with type-2 fuzzy weights (see Fig. 3.32) is described as follows:

Layer 0: Inputs.

$$x = [x_1, x_2, \cdots, x_n] \tag{3.15}$$

Layer 1: Interval type-2 fuzzy weights for the hidden layer [22].

$$\tilde{w} = [\underline{w}, \bar{w}] \tag{3.16}$$

where:

$$\bar{w} = \bar{w}\left(\bar{f}^1, \ldots, \bar{f}^L, \underline{f}^{L+1}, \ldots, \underline{f}^M, w_1^1, \ldots, w_1^M\right) = \frac{\sum_{k=1}^{L} \bar{f}^k \cdot w_1^k + \sum_{k=L+1}^{M} \underline{f}^k \cdot w_1^k}{\sum_{k=1}^{L} \bar{f}^k + \sum_{k=L+1}^{M} \underline{f}^k} \tag{3.17}$$

$$\underline{w} = \bar{w}\left(\underline{f}^1, \ldots, \underline{f}^R, \bar{f}^{R+1}, \ldots, \bar{f}^M, w_r^1, \ldots, w_r^M\right) = \frac{\sum_{k=1}^{R} \underline{f}^k \cdot w_r^k + \sum_{k=R+1}^{M} \bar{f}^k \cdot w_r^k}{\sum_{k=1}^{R} \underline{f}^k + \sum_{k=R+1}^{M} \bar{f}^k} \tag{3.18}$$

where w_1^k and w_r^k are the consequent left-right firing points, $\bar{f}^k$ and $\underline{f}^k$ are the firing for the rules, L and R are the switch points [22].

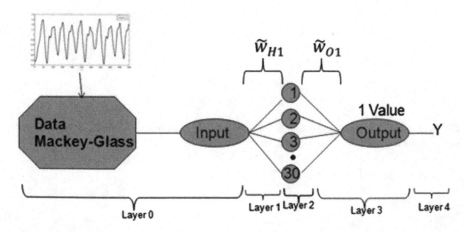

Fig. 3.32 Neural network architecture with type-2 fuzzy weights for Mackey-Glass

Layer 2: Hidden neuron with interval type-2 fuzzy weights.

$$\mathrm{Net} = \sum_{i=1}^{n} x_i \widetilde{w_i} \qquad (3.19)$$

Layer 3: Output neuron with interval type-2 fuzzy weights.

$$\mathrm{Out} = \sum_{i=1}^{n} y_i \widetilde{w_i} \qquad (3.20)$$

Layer 4: Obtain a single output of the neural network.

The adaptation of type-2 fuzzy weights is based on the back-propagation algorithm in the same way for the type-1 fuzzy weights.

We used 2 similar type-2 fuzzy inference systems to obtain the type-2 fuzzy weights in the hidden and output layer for the neural network.

For obtaining the type-2 fuzzy inference systems an extension of the membership functions of the type-1 fuzzy inference systems was applied.

We increase or decrease the values for the triangular membership functions with a variable epsilon in terms of percentage to obtain the footprint of uncertainty (FOU). These are applied in the type-2 fuzzy inference systems used in the neural network with type-2 fuzzy weights [23].

We present, for example, the membership functions obtained with an epsilon of ± 1 % for the inputs and output the type-2 fuzzy inference systems. The first type-2 fuzzy system consists of two inputs: the current weight in the actual epoch and the

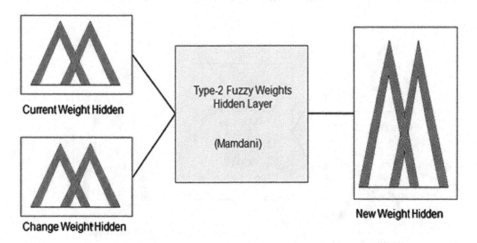

Fig. 3.33 Type-2 fuzzy inference system used in the hidden layer for the neural network with type-2 fuzzy weights (triangular MF) for Mackey-Glass

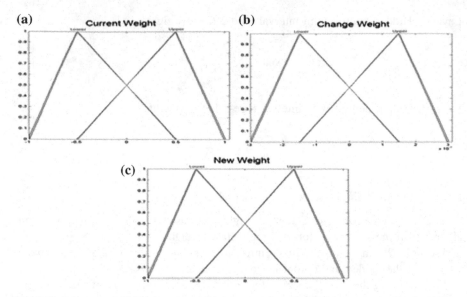

Fig. 3.34 Inputs (**a** and **b**) and output (**c**) of the type-2 fuzzy inference system used in the hidden layer for the neural network with type-2 fuzzy weights (triangular MF) for Mackey-Glass

change of the weight for the next epoch, and one output: the new weight for the next epoch (see Fig. 3.33).

The input of the current weight has two triangular membership functions with range of −1 to 1. The input of change of the weight has two triangular membership functions with range of −0.003 to 0.003. The output of the new weight has two triangular membership functions with range of −1 to 1 (see Fig. 3.34).

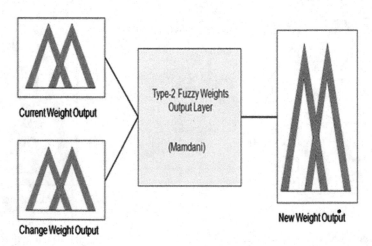

Fig. 3.35 Type-2 fuzzy inference system used in the output layer for the neural network with type-2 fuzzy weights (triangular MF) for Mackey-Glass

We used the same six rules before presented for the type-2 fuzzy inference system for the hidden layer (see Fig. 3.29).

The second type-2 fuzzy system consists of two inputs: the current weight in the actual epoch and the change of the weight for the next epoch, and one output: the new weight for the next epoch (see Fig. 3.35).

The input of the current weight has two triangular membership functions with range of -1 to 1. The input of the change of the weight has two triangular membership functions with range of -0.01 to 0.01. The output of the new weight has two triangular membership functions with range of -1 to 1 (see Fig. 3.36).

We used the same six rules before presented for the type-2 fuzzy inference system for the output layer (see Fig. 3.29).

The experiments were performed in time-series prediction, specifically for the Mackey-Glass time series (for $\tau = 17$) that behaves chaotically and it is a benchmark used in many studies.

We used the gradient descent backpropagation algorithm with adaptive learning rate for the experiments.

The neural networks handle type-1 and type-2 fuzzy weights in each one of their hidden layers and output layer [21, 24].

In each hidden layer and output of each network we are using a type-1 fuzzy inference system or type-2 fuzzy inference system to obtain new weights in each epoch of the network [18–20].

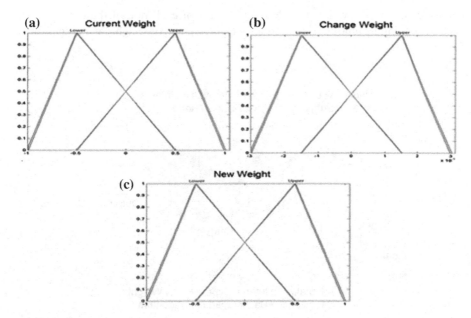

Fig. 3.36 Inputs (**a** and **b**) and output (**c**) of the type-2 fuzzy inference system used in the output layer for the neural network with type-2 fuzzy weights (triangular MF) for Mackey-Glass

3.6   Neural Network with Lower and Upper Type-2 Fuzzy Weights Using the Backpropagation Learning Method

The proposed method performs the update of the weights working with interval type-2 fuzzy weights, this method use the left type-2 fuzzy weight, the right type-2 fuzzy weights and the interval type-2 fuzzy weights for obtaining the outputs taking into account the possible change in the way we work internally in the neuron, and the adaptation of the weights given in this way (Fig. 3.37).

The proposed neural network architecture with interval type-2 fuzzy weights (NNT2FWLU) (see Fig. 3.38) is described as follows:

Phase 0: Inputs.

$$x = [x_1, x_2, \ldots, x_n] \tag{3.21}$$

Phase 1: Interval type-2 fuzzy weights for the connection between the input and the hidden layer of the neural network.

$$\tilde{w} = [\underline{w}, \bar{w}] \tag{3.22}$$

where $[\underline{w}, \bar{w}]$ are the weights of the consequents of each rule of the type-2 fuzzy system with inputs (current type-2 fuzzy weight, change of weight) and output (new fuzzy weight).

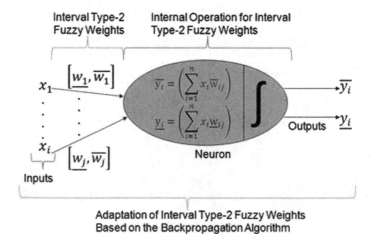

Fig. 3.37 Schematic of each neuron with the proposed management of weights using interval type 2 fuzzy sets (lower and upper) for Mackey-Glass

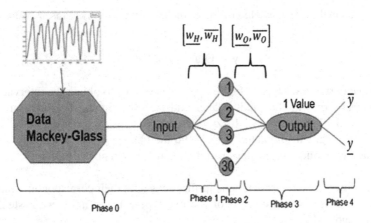

Fig. 3.38 Proposed neural network architecture with interval type-2 fuzzy weights (lower and upper) for Mackey-Glass

Phase 2: Equations of the calculations in the hidden neurons using interval type-2 fuzzy weights.

$$\underline{\text{Net}} = \sum_{i=1}^{n} x_i \underline{w_i} \qquad (3.23)$$

$$\overline{\text{Net}} = \sum_{i=1}^{n} x_i \overline{w_i} \qquad (3.24)$$

Phase 3: Equations of the calculations in the outputs neurons using interval type-2 fuzzy weights.

$$\underline{\text{Out}} = \sum_{i=1}^{n} y_i \underline{w_i} \qquad (3.25)$$

$$\overline{\text{Out}} = \sum_{i=1}^{n} y_i \overline{w_i} \qquad (3.26)$$

Phase 4: Obtain the upper and lower output of the neural network (Figs. 3.39 and 3.40).

$$\underline{y} = \underline{Out} \quad \bar{y} = \overline{Out} \tag{3.27}$$

We used two similar type-2 fuzzy inference systems to obtain the interval type-2 fuzzy weights, one in the hidden layer and the other in the output layer for the neural network. The two type-2 fuzzy inference systems used consist for two inputs (the current weight in the actual epoch and the change of the weight for the next epoch) and one output (the new weight for the next epoch) (see Figs. 3.39 and 3.41).

We used two triangular membership functions with their corresponding range for delimiting the inputs and outputs of the two type-2 fuzzy inference systems (see Figs. 3.40 and 3.42).

We obtain the type-2 fuzzy inference systems incrementing and decrementing the values for the triangular membership functions with a variable epsilon in terms of percentage, we use to obtain the footprint of uncertainty (FOU) for the type-2 fuzzy inference systems used in the neural network with type-2 fuzzy weights (the triangular membership functions used are of an previous research) [23]. We present, for example, the membership functions for the inputs and output of the type-2 fuzzy inference systems obtained with an epsilon of ± 2 %.

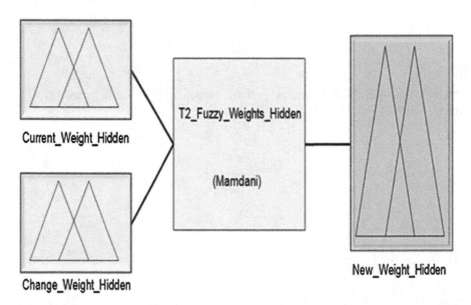

Fig. 3.39 Structure of the type-2 fuzzy inference system that used for obtain the type-2 fuzzy weights in the hidden layer in NNT2FWLU for Mackey-Glass

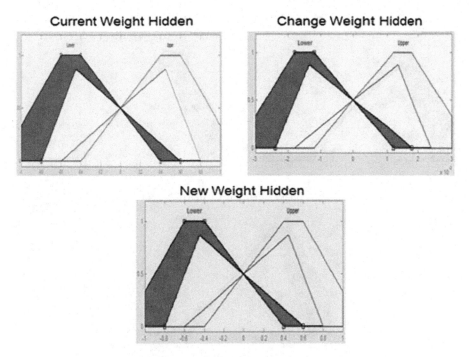

Fig. 3.40 Inputs and outputs of the type-2 fuzzy inference system that used for obtain the type-2 fuzzy weights in the hidden layer in NNT2FWLU for Mackey-Glass

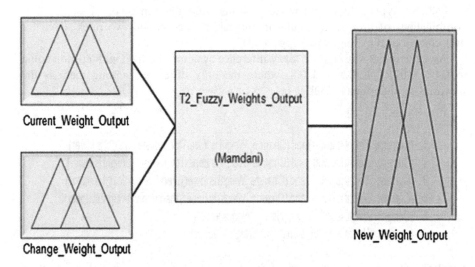

Fig. 3.41 Structure of the type-2 fuzzy inference system that used for obtain the type-2 fuzzy weights in the output layer in NNT2FWLU for Mackey-Glass

Current Weight Output

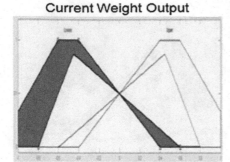

Change Weight Output

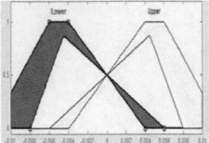

New Weight Output

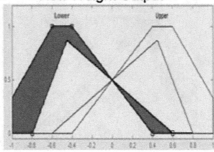

Fig. 3.42 Inputs and outputs of the type-2 fuzzy inference system that used for obtain the type-2 fuzzy weights in the output layer in NNT2FWLU for Mackey-Glass

The two type-2 fuzzy inference systems used the same six rules, the four combinations of the two triangular membership function and two rules added for null change of the weight (see Fig. 3.43).

We experiment with type-2 fuzzy inference systems obtained with epsilon equal to 0.01, 0.07, 0.1, 0.2 and 0.8, where the only difference among them is the footprint of uncertainty (FOU).

1. (Current_Weight is lower) and (Change_Weight is lower) then (New_Weight is lower)
2. (Current_Weight is lower) and (Change_Weight is upper) then (New_Weight is lower)
3. (Current_Weight is upper) and (Change_Weight is lower) then (New_Weight is upper)
4. (Current_Weight is upper) and (Change_Weight is upper) then (New_Weight is upper)
5. (Current_Weight is lower) then (New_Weight is lower)
6. (Current_Weight is upper) then (New_Weight is upper)

Fig. 3.43 Rules of the two type-2 fuzzy inference systems that used for obtain the type-2 fuzzy weights in the hidden and output layer in NNT2FWLU for Mackey-Glass

3.7 Optimization of Type-2 Fuzzy Weights for the Neural Network Using Genetic Algorithm and Particle Swarm Optimization (NNT2FWGA-PSO)

The proposed method will work with interval type-2 fuzzy weights, taking into account the possible change in the way we work internally in the neuron, the optimization with genetic algorithm and particle swarm optimization is applied in the type-2 fuzzy inference systems used in each layer for obtaining the type-2 fuzzy weights and finally obtained the output (Fig. 3.44).

The proposed neural network architecture with interval type-2 fuzzy weights (see Fig. 3.45) is described as follows:

Layer 0: Inputs.

$$\mathbf{x} = [x_1, x_2, \ldots, x_n] \tag{3.28}$$

Layer 1: Interval type-2 fuzzy weights for the hidden layer of each neural network.

$$\tilde{w} = [\underline{w}, \bar{w}] \tag{3.29}$$

where $[\underline{w}, \bar{w}]$ are the weights of the consequents of each rule of the type-2 fuzzy system with inputs (current fuzzy weight, change of weight) and output (new fuzzy weight).

Fig. 3.44 Schematic of the proposed optimization for the management of interval type 2 fuzzy weights for input of each neuron using GA and PSO

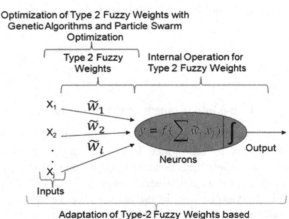

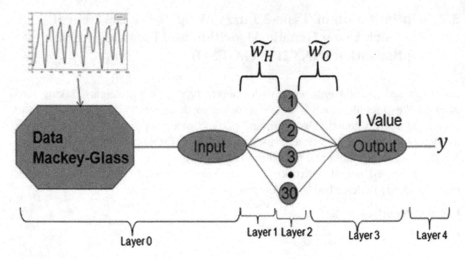

Fig. 3.45 Proposed neural network architecture with interval type-2 fuzzy weights for NNT2FWGA-PSO

Layer 2: Hidden neuron with type-2 fuzzy weights.

$$\text{Net} = \sum_{i=1}^{n} x_i \widetilde{w_i} \qquad (3.30)$$

Layer 3: Output neuron with type-2 fuzzy weights.

$$\text{Out} = \sum_{i=1}^{n} y_i \widetilde{w_i} \qquad (3.31)$$

Layer 4: Obtain one output of the neural network.

We considered neural network architecture with 1 neuron in the output layer and 30 neurons in the hidden layer. This neural network use interval type-2 fuzzy weights in the connections between the input neurons and the hidden neurons, and in the connections between hidden neurons and the output neuron. In the hidden layer and output layer of the network we are updating the weights using a interval type-2 fuzzy inference system that obtains the new weights in each epoch of the network on base at the backpropagation algorithm [18–21].

We used two similar type-2 fuzzy inference systems to obtain the type-2 fuzzy weights in the hidden and output layer for the neural network.

For obtaining the type-2 fuzzy inference systems we used two bio-inspired methods: genetic algorithm (GA) and particle swarm optimization (PSO) (Fig. 3.46).

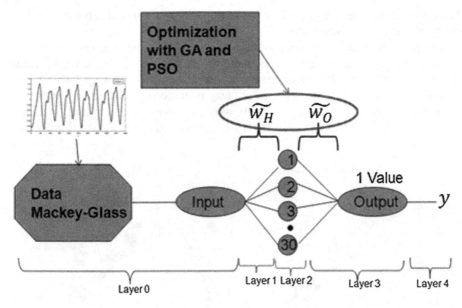

Fig. 3.46 Proposed optimization of the neural network architecture with type-2 fuzzy weights for NNT2FWGA-PSO

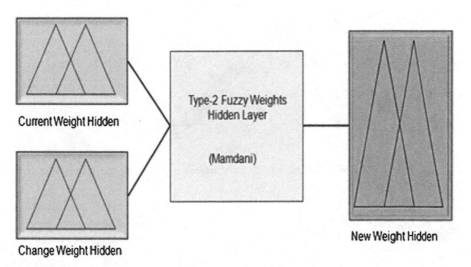

Fig. 3.47 Structure of the used type-2 fuzzy inference system in the hidden layer for NNT2FWGA-PSO

The optimization was not applied for the architecture of the neural network only for the type-2 fuzzy inference systems.

The two type-2 fuzzy inference systems consist of the same structure: two inputs (the current weight in the actual epoch and the change of the weight for the next epoch) and one output (the new weight for the next epoch) (see Fig. 3.47).

The inputs and outputs for the type-2 fuzzy inference systems without optimize are delimited with two triangular membership functions with their corresponding range (see Figs. 3.48 and 3.49). The rules for the both type-2 fuzzy inference

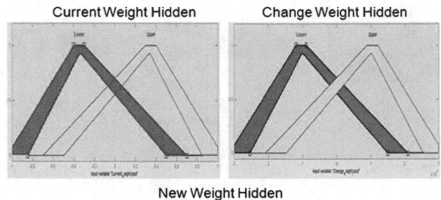

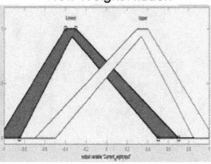

Fig. 3.48 Inputs and outputs of the type-2 fuzzy inference system for the hidden layer for NNT2FWGA-PSO

Current Weight Output

Change Weight Output

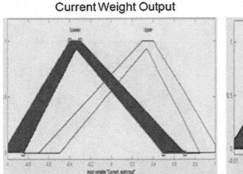
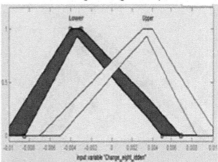

New Weight Output

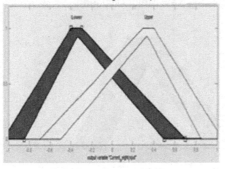

Fig. 3.49 Inputs and outputs of the type-2 fuzzy inference system for the output layer for NNT2FWGA-PSO

---

1. (Current_Weight is lower) and (Change_Weight is lower) then (New_Weight is lower)
2. (Current_Weight is lower) and (Change_Weight is upper) then (New_Weight is lower)
3. (Current_Weight is upper) and (Change_Weight is lower) then (New_Weight is upper)
4. (Current_Weight is upper) and (Change_Weight is upper) then (New_Weight is upper)
5. (Current_Weight is lower) then (New_Weight is lower)
6. (Current_Weight is upper) then (New_Weight is upper)

Fig. 3.50 Rules of the type-2 fuzzy inference systems used in the hidden and output layer for NNT2FWGA-PSO

systems are the same, and these were created for the combination of the inputs and outputs and two rules when the change of weight was zero (see Fig. 3.50).

The two optimized type-2 fuzzy inference systems optimized with genetic algorithm have the same structure and the same rules that have the type-2 fuzzy inference systems without optimizing (see Figs. 3.51 and 3.52).

Current Weight Hidden

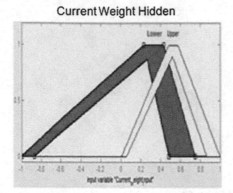

Change Weight Hidden

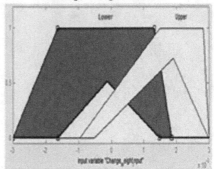

New Weight Hidden

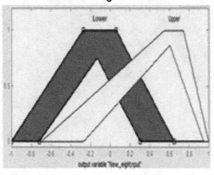

Fig. 3.51 Inputs and outputs of the type-2 fuzzy inference system optimized with genetic algorithm for the hidden layer for NNT2FWGA

The parameters of the genetic algorithm using to optimize the two type-2 fuzzy inference systems are shown on Table 3.2. The chromosomes are used for each membership function each ten gens of the two inputs and one output of the two type-2 fuzzy inference systems.

The two type-2 fuzzy inference systems optimized with Particle swarm optimization are the same structure and the same rules that have the type-2 fuzzy inference systems without optimizing (see Figs. 3.53 and 3.54).

The parameters of the particle swarm optimization used for optimizing the two type-2 fuzzy inference systems are shown on Table 3.3. The particles are used for each membership function each ten dimensions of the two inputs and one output of the two type-2 fuzzy inference systems.

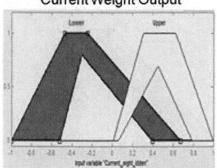
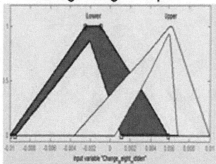

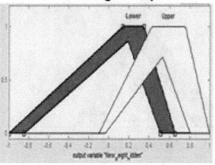

Fig. 3.52 Inputs and outputs of the type-2 fuzzy inference system optimized with genetic algorithm for the output layer for NNT2FWGA

Table 3.2 Parameters of the genetic algorithm used for the optimization of the two type-2 fuzzy inference systems for NNT2FWGA

Individuals	100
Gens	60 gens (real)
Generations	65
Assign fitness	Ranking
Selection	Stochastic universal sampling
Crossover	Single-point (0.8)
Mutation	0.4

Current Weight Hidden

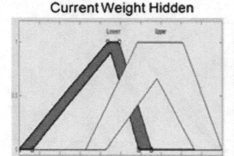

Change Weight Hidden

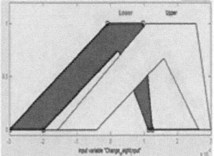

New Weight Hidden

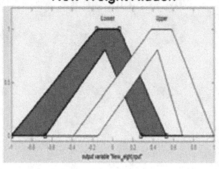

Fig. 3.53 Inputs and outputs of the type-2 fuzzy inference system optimized with particle swarm optimization for the hidden layer for NNT2FWPSO

3.8   Neural Network with Fuzzy Weights Using Type-1 and Type-2 Fuzzy Learning with Gaussian Membership Functions

The objective of this work is to use type-1 and interval type-2 fuzzy sets to generalize the backpropagation algorithm to allow the neural network to handle data with uncertainty. The Mackey-Glass time series (for $\tau = 17$) is utilized for testing the proposed approach.

The proposed method performs the updating of the weights working with interval type-2 fuzzy weights, this method uses two type-1 and two type-2 inference systems with Gaussian membership functions for obtaining the type-1 and interval type-2 fuzzy weights using in the neural network.

Current Weight Output

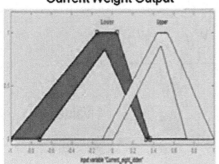

Change Weight Output

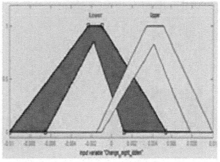

New Weight Output

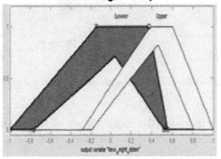

Fig. 3.54 Inputs and outputs of the type-2 fuzzy inference system optimized with particle swarm optimization for the output layer for NNT2FWPSO

Table 3.3 Parameters of the particle swarm optimization used for the optimization of the two type-2 fuzzy inference systems		
	Particles	100
	Dimensions	60 (real)
	Iterations	65
	Inertia weight (ω)	Linear decrement (0.88–0)
	Constriction (C)	Linear increment (0.01–0.9)
	r1, r2	Random
	c1	Linear decrement (2–0.5)
	c2	Linear increment (0.5–2)

The proposed neural network architecture with type-1 fuzzy weights (see Fig. 3.55) is described as follows:

Phase 0: Inputs.

$$x = [x_1, x_2, \ldots, x_n] \tag{3.32}$$

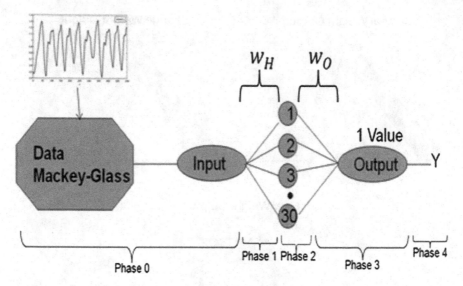

Fig. 3.55 Proposed neural network architecture with type-1 fuzzy weights (Gaussian MF)

Phase 1: Type-1 fuzzy weights for the hidden layer of the neural network.

$$w_{ij} \qquad (3.33)$$

Phase 2: Equations of the calculations in the hidden neurons using type-1 fuzzy weights.

$$Net = \sum_{i=1}^{n} x_i w_i \qquad (3.34)$$

Phase 3: Equations of the calculations in the outputs neurons using type-1 fuzzy weights.

$$Out = \sum_{i=1}^{n} y_i w_i \qquad (3.35)$$

Phase 4: Obtain one output of the neural network.

We considered neural network architecture with 1 neuron in the output layer and 30 neurons in the hidden layer.

This neural network used two type-1 fuzzy inference systems, one in the connections between the input neurons and the hidden neurons, and the other in the

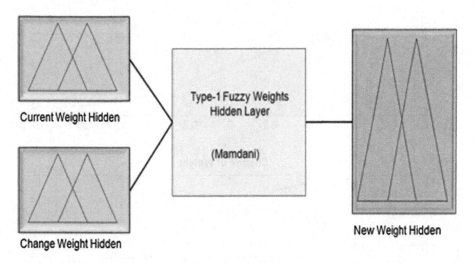

Fig. 3.56 Structure of the two type-1 fuzzy inference systems (Gaussian MF) that were used to obtain the type-1 fuzzy weights in the hidden and output layer

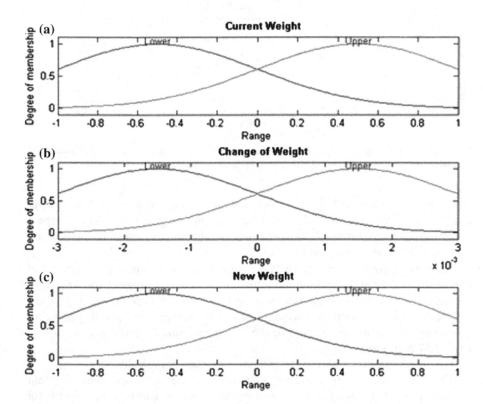

Fig. 3.57 Inputs (**a** and **b**) and output (**c**) of the type-1 fuzzy inference systems (Gaussian MF) that were used to obtain the type-1 fuzzy weights in the hidden layer

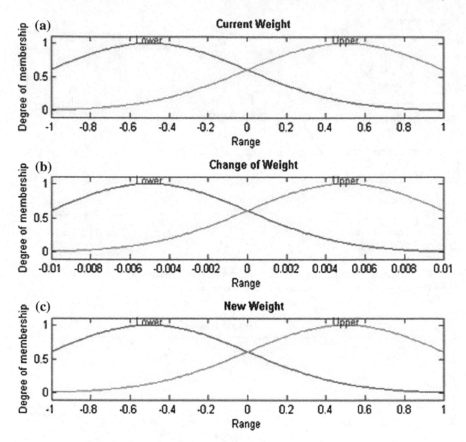

Fig. 3.58 Inputs (**a** and **b**) and output (**c**) of the type-1 fuzzy inference systems (Gaussian MF) that were used to obtain the type-1 fuzzy weights in the output layer

connections between the hidden neurons and the output neuron. In the hidden layer and output layer of the network we are updating the weights using the two type-1 fuzzy inference system that obtains the new weights in each epoch of the network on base at the backpropagation algorithm.

The two type-2 fuzzy inference systems have the same structure and consist of two inputs (the current weight in the actual epoch and the change of the weight for the next epoch) and one output (the new weight for the next epoch) (see Fig. 3.56).

We used two Gaussian membership functions with their corresponding range for delimiting the inputs and outputs of the two type-1 fuzzy inference systems (see Figs. 3.57 and 3.58).

We obtain the two type-1 fuzzy inference systems empirically.

The two type-1 fuzzy inference systems used the same six rules, the four combinations of the two Gaussian membership function and two rules added for null change of the weight (see Fig. 3.59).

1. (Current_Weight is lower) and (Change_Weight is lower) then (New_Weight is lower)
2. (Current_Weight is lower) and (Change_Weight is upper) then (New_Weight is lower)
3. (Current_Weight is upper) and (Change_Weight is lower) then (New_Weight is upper)
4. (Current_Weight is upper) and (Change_Weight is upper) then (New_Weight is upper)
5. (Current_Weight is lower) then (New_Weight is lower)
6. (Current_Weight is upper) then (New_Weight is upper)

Fig. 3.59 Rules of the type-2 fuzzy inference systems (Gaussian MF) used in the hidden and output layer

The proposed neural network architecture with interval type-2 fuzzy weights (see Fig. 3.60) is described as follows:

Layer 0: Inputs.

$$x = [x_1, x_2, \ldots, x_n] \tag{3.36}$$

Layer 1: Interval type-2 fuzzy weights for the connection between the input and the hidden layer of the neural network.

$$\tilde{w}_{ij} = \left[\bar{w}_{ij}, \underline{w}_{ij} \right] \tag{3.37}$$

where $\tilde{w}_{ij}$ are the weights of the consequents of each rule of the type-2 fuzzy system with inputs (current fuzzy weight, change of weight) and output (new fuzzy weight).

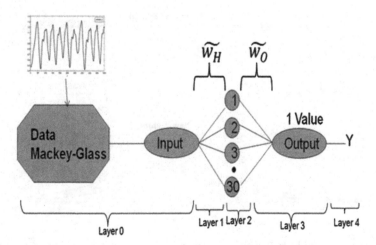

Fig. 3.60 Proposed neural network architecture with interval type-2 fuzzy weights (Gaussian MF)

Layer 2: Equations of the calculations in the hidden neurons using interval type-2 fuzzy weights.

$$Net = \sum_{i=1}^{n} x_i \tilde{w}_{ij}$$
(3.38)

Layer 3: Equations of the calculations in the output neurons using interval type-2 fuzzy weights.

$$Out = \sum_{i=1}^{n} y_i \tilde{w}_{ij}$$
(3.39)

Layer 4: Obtain a single output of the neural network.

We used two type-2 fuzzy inference systems to obtain the type-2 fuzzy weights and work in the same way like with the type-1 fuzzy weights.

The structure and rules (see Fig. 3.55) of the two type-2 fuzzy inference systems are the same of the type-1 fuzzy inference systems; the difference is in the member-ships functions, Gaussian membership functions for type-2.

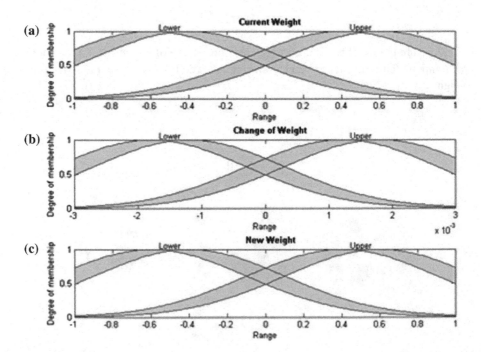

Fig. 3.61 Inputs (**a** and **b**) and output (**c**) of the type-1 fuzzy inference systems (Gaussian MF) that were used to obtain the interval in the hidden layer

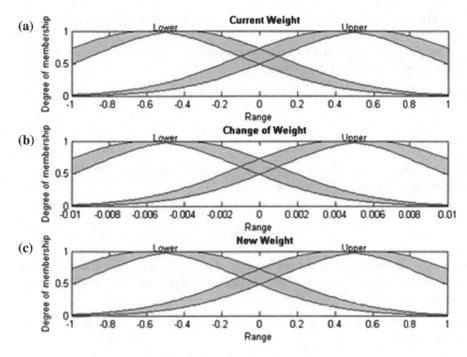

Fig. 3.62 Inputs (**a** and **b**) and output (**c**) of the type-1 fuzzy inference systems (Gaussian MF) that were used to obtain the interval type-2 fuzzy weights in the output layer

We used two Gaussian membership functions with their corresponding range for delimiting the inputs and outputs of the two type-2 fuzzy inference systems (see Figs. 3.61 and 3.62).

We obtain the type-2 fuzzy inference systems incrementing and decrementing 20 % the values of the centers of the Gaussian membership functions and the same standard deviation of the type-1 Gaussians membership functions, we use this method to obtain the footprint of uncertainty (FOU) for the type-2 fuzzy inference systems used in the neural network with type-2 fuzzy weights.

3.9 Comparison of Neural Network Performance Under Different Membership Functions in the Type-2 Fuzzy Weights (CMFNNT2FW)

The objective of this work is to use interval type-2 fuzzy sets to generalize the back-propagation algorithm to allow the neural network to handle data with uncertainty, and a comparison of the type of membership functions in the type-2 fuzzy systems, which can be triangular, Gaussian, trapezoidal and generalized bell.

The Mackey-Glass time series (for $\tau = 17$) is utilized for testing the proposed approach.

The proposed neural network architecture with interval type-2 fuzzy weights (see Fig. 3.63) is described as follows:

Phase 0: Inputs.

$$x = [x_1, x_2, \ldots, x_n] \tag{3.40}$$

Phase 1: Interval type-2 fuzzy weights for the connection between the input and the hidden layer of the neural network.

$$\tilde{w}_{ij} = \left[\bar{w}_{ij}, \underline{w}_{ij}\right] \tag{3.41}$$

where $\tilde{w}_{ij}$ are the weights of the consequents of each rule of the type-2 fuzzy system with inputs (current fuzzy weight, change of weight) and output (new fuzzy weight).

Phase 2: Equations of the calculations in the hidden neurons using interval type-2 fuzzy weights.

$$\text{Net} = \sum_{i=1}^{n} x_i \tilde{w}_{ij} \tag{3.42}$$

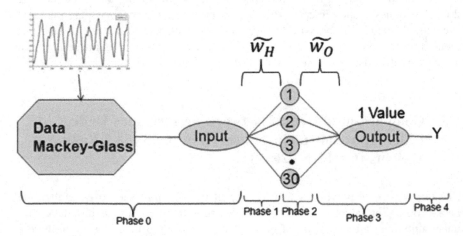

Fig. 3.63 Proposed neural network architecture with interval type-2 fuzzy weights (CMFNNT2FW)

Phase 3: Equations of the calculations in the output neurons using interval type-2 fuzzy weights.

$$\text{Out} = \sum_{i=1}^{n} y_i \tilde{w}_{ij} \tag{3.43}$$

Phase 4: Obtain a single output of the neural network.

We considered neural network architecture with 1 neuron in the output layer and 30 neurons in the hidden layer.

This neural network uses two type-2 fuzzy inference systems, one in the connections between the input neurons and the hidden neurons, and the other in the connections between the hidden neurons and the output neuron. In the hidden layer and output layer of the network we are updating the weights using the two type-2 fuzzy inference system that obtains the new weights in each epoch of the network on base at the back-propagation algorithm.

The two type-2 fuzzy inference systems have the same structure and consist of two inputs (the current weight in the actual epoch and the change of the weight for the next epoch) and one output (the new weight for the next epoch) (see Fig. 3.64).

We used two membership functions with their corresponding range for delimiting the inputs and outputs of the two type-2 fuzzy inference systems.

We used triangular, trapezoidal, Gaussian and generalized bell membership functions (for example, see Figs. 3.65 and 3.66).

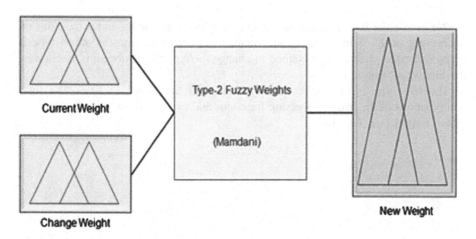

Fig. 3.64 Structure of the two interval type-2 fuzzy inference systems that were used to obtain the interval type-2 fuzzy weights in the hidden and output layer for CMFNNT2FW

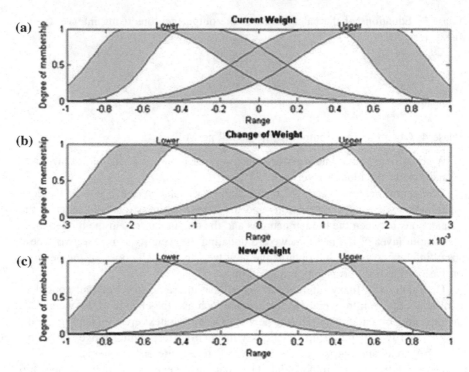

Fig. 3.65 Inputs (**a** and **b**) and output (**c**) of the type-1 fuzzy inference systems that were used to obtain the type-2 fuzzy weights in the hidden layer for CMFNNT2FW

We obtain the two type-2 fuzzy inference systems manually and a footprint of uncertainty of 15 % in the triangular membership functions and the others type-2 fuzzy inference systems are obtained to change the type of membership function in the inputs and output.

The two type-2 fuzzy inference systems used the same six rules, the four combinations of the two membership functions and two rules added for null change of the weight (see Fig. 3.67).

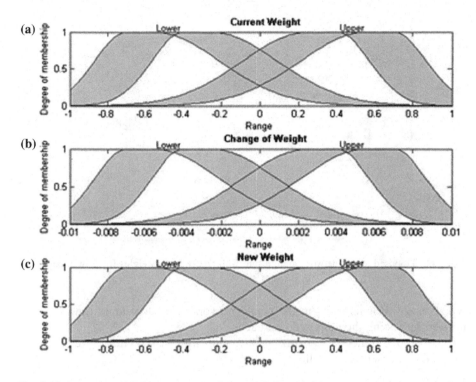

Fig. 3.66 Inputs (**a** and **b**) and output (**c**) of the type-2 fuzzy inference systems that were used to obtain the type-1 fuzzy weights in the output layer for CMFNNT2FW

1. (Current_Weight is lower) and (Change_Weight is lower) then (New_Weight is lower)
2. (Current_Weight is lower) and (Change_Weight is upper) then (New_Weight is lower)
3. (Current_Weight is upper) and (Change_Weight is lower) then (New_Weight is upper)
4. (Current_Weight is upper) and (Change_Weight is upper) then (New_Weight is upper)
5. (Current_Weight is lower) then (New_Weight is lower)
6. (Current_Weight is upper) then (New_Weight is upper)

Fig. 3.67 Rules of the type-2 fuzzy inference systems used in the hidden and output layer for CMFNNT2FW

3.10 Neural Network with Fuzzy Weights Using Type-1 and Type-2 Fuzzy Learning for Dow-Jones Time Series

The proposed method performs the updating of the weights working with interval type-2 fuzzy weights, this method uses two type-1 and two type-2 inference systems with Gaussian membership functions for obtaining the type-1 and interval

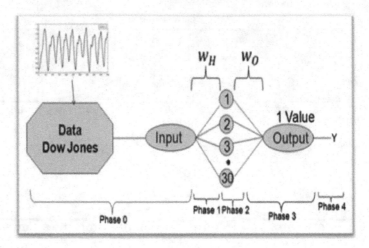

Fig. 3.68 Proposed neural network architecture with type-1 fuzzy weights for Dow-Jones

type-2 fuzzy weights using in the neural network. The Dow-Jones time series is utilized for testing the proposed approach.

The proposed neural network architecture with type-1 fuzzy weights (see Fig. 3.68) is described as follows:

Phase 0: Inputs.

$$x = [x_1, x_2, \ldots, x_n] \tag{3.44}$$

Phase 1: Type-1 fuzzy weights for the hidden layer of the neural network.

$$w_{ij} \tag{3.45}$$

Phase 2: Equations of the calculations in the hidden neurons using type-1 fuzzy weights..

$$Net = \sum_{i=1}^{n} x_i w_i \tag{3.46}$$

Phase 3: Equations of the calculations in the outputs neurons using type-1 fuzzy weights.

$$\text{Out} = \sum_{i=1}^{n} y_i w_i \qquad (3.47)$$

Phase 4: Obtain one output of the neural network.

This neural network used two type-1 fuzzy inference systems, one in the connections between the input neurons and the hidden neurons, and the other in the connections between the hidden neurons and the output neuron. In the hidden layer and output layer of the network we are updating the weights using the two type-1 fuzzy inference system that obtains the new weights in each epoch of the network on base at the back-propagation algorithm.

The two type-2 fuzzy inference systems have the same structure and consist of two inputs (the current weight in the actual epoch and the change of the weight for the next epoch) and one output (the new weight for the next epoch) (see Fig. 3.69).

We used two Gaussian membership functions with their corresponding range for delimiting the inputs and outputs of the two type-1 fuzzy inference systems (see Figs. 3.70 and 3.71).

We obtain the two type-1 fuzzy inference systems empirically.

The two type-1 fuzzy inference systems used the same six rules, the four combinations of the two Gaussian membership function and two rules added for null change of the weight (see Fig. 3.72).

The proposed neural network architecture with interval type-2 fuzzy weights (see Fig. 3.73) is described as follows:

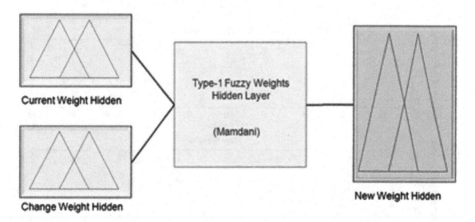

Fig. 3.69 Structure of the two type-1 fuzzy inference systems that were used to obtain the type-1 fuzzy weights in the hidden and output layer for Dow-Jones

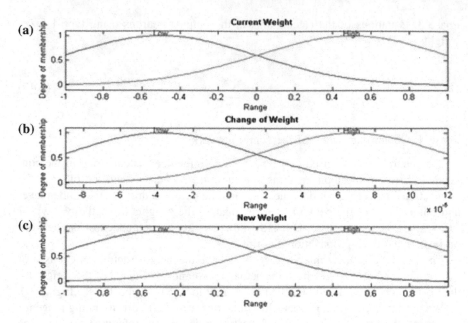

Fig. 3.70 Inputs (**a** and **b**) and output (**c**) of the type-1 fuzzy inference systems that were used to obtain the type-1 fuzzy weights in the hidden layer

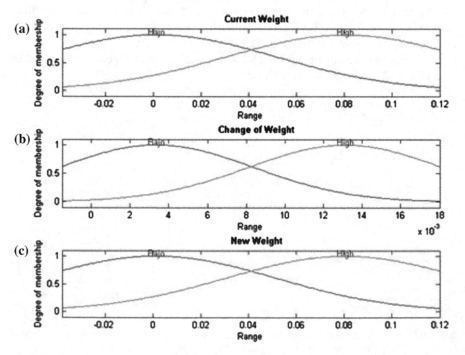

Fig. 3.71 Inputs (**a** and **b**) and output (**c**) of the type-1 fuzzy inference systems that were used to obtain the type-1 fuzzy weights in the output layer for Dow-Jones

> 1. (Current_Weight is lower) and (Change_Weight is lower) then (New_Weight is lower)
> 2. (Current_Weight is lower) and (Change_Weight is upper) then (New_Weight is lower)
> 3. (Current_Weight is upper) and (Change_Weight is lower) then (New_Weight is upper)
> 4. (Current_Weight is upper) and (Change_Weight is upper) then (New_Weight is upper)
> 5. (Current_Weight is lower) then (New_Weight is lower)
> 6. (Current_Weight is upper) then (New_Weight is upper)

Fig. 3.72 Rules of the type-2 fuzzy inference systems used in the hidden and output layer for Dow-Jones

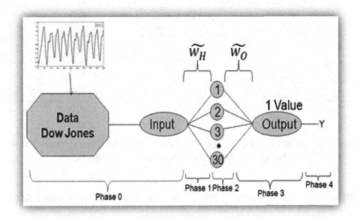

Fig. 3.73 Proposed neural network architecture with interval type-2 fuzzy weights for Dow-Jones

Phase 0: Inputs.

$$x = [x_1, x_2, \ldots, x_n] \tag{3.48}$$

Phase 1: Interval type-2 fuzzy weights for the connection between the input and the hidden layer of the neural network.

$$\tilde{w}_{ij} = \left[\overline{w}_{ij}, \underline{w}_{ij} \right] \tag{3.49}$$

where $\tilde{w}_{ij}$ are the weights of the consequents of each rule of the type-2 fuzzy system with inputs (current fuzzy weight, change of weight) and output (new fuzzy weight).

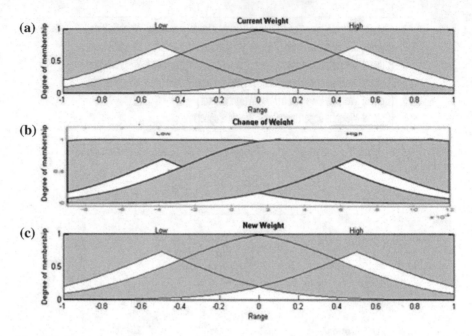

Fig. 3.74 Inputs (**a** and **b**) and output (**c**) of the type-1 fuzzy inference systems that were used to obtain the interval type-2 fuzzy weights in the hidden layer for Dow-Jones

Phase 2: Equations of the calculations in the hidden neurons using interval type-2 fuzzy weights.

$$\text{Net} = \sum_{i=1}^{n} x_i \tilde{w}_{ij} \tag{3.50}$$

Phase 3: Equations of the calculations in the output neurons using interval type-2 fuzzy weights.

$$\text{Out} = \sum_{i=1}^{n} y_i \tilde{w}_{ij} \tag{3.51}$$

Phase 4: Obtain a single output of the neural network.

We used two type-2 fuzzy inference systems to obtain the type-2 fuzzy weights and work in the same way like with the type-1 fuzzy weights.

The structure and rules (see Figs. 3.70 and 3.71) of the two type-2 fuzzy inference systems are the same of the type-1 fuzzy inference systems; the difference is in the member-ships functions, Gaussian membership functions for type-2.

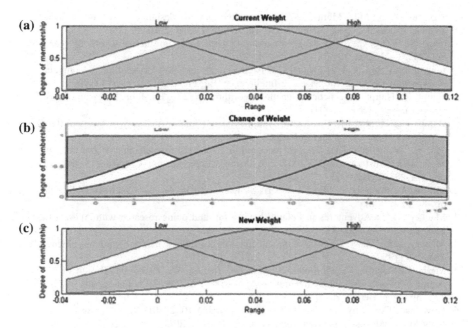

Fig. 3.75 Inputs (**a** and **b**) and output (**c**) of the type-1 fuzzy inference systems that were used to obtain the interval type-2 fuzzy weights in the output layer for Dow-Jones

We used two Gaussian membership functions with their corresponding range for delimiting the inputs and outputs of the two type-2 fuzzy inference systems (see Figs. 3.74 and 3.75).

We obtain the type-2 fuzzy inference systems incrementing and decrementing 20 % the values of the centers of the Gaussian membership functions and the same standard deviation of the type-1 Gaussians membership functions, we use this method to obtain the footprint of uncertainty (FOU) for the type-2 fuzzy inference systems used in the neural network with type-2 fuzzy weights.

References

1. Pedrycz, W.: Granular Computing: Analysis and Design of Intelligent Systems. CRC Press/Francis Taylor, Boca Raton (2013)
2. Tung, S.W., Quek, C., Guan, C.: eT2FIS: an evolving type-2 neural fuzzy inference system. Inf. Sci. **220**, 124–148 (2013)
3. Zarandi, M.H.F., Torshizi, A.D., Turksen, I.B., Rezaee, B.: A new indirect approach to the type-2 fuzzy systems modeling and design. Inf. Sci. **232**, 346–365 (2013)
4. Zhai, D., Mendel, J.: Uncertainty measures for general type-2 fuzzy sets. Inf. Sci. **181**(3), 503–518 (2011)
5. Biglarbegian, M., Melek, W., Mendel, J.: On the robustness of type-1 and interval type-2 fuzzy logic systems in modeling. Inf. Sci. **181**(7), 1325–1347 (2011)

6. Monirul Islam, Md., Murase, K.: A new algorithm to design compact two-hidden-layer artificial neural networks. Neural Netw. **14**(9), 1265–1278 (2001)
7. Chen, S., Wang, C.: Fuzzy decision making systems based on interval type-2 fuzzy sets. Inf. Sci. **242**, 1–21 (2013)
8. Jang, J.S.R., Sun, C.T., Mizutani, E.: Neuro-Fuzzy and Soft Computing: a Computational Approach to Learning and Machine Intelligence. Ed. Prentice Hall, p. 614 (1997)
9. Haupt, R., Haupt, S.: Practical Genetic Algorithms, p. 272. Ed. John Wiley and Sons, Inc., Hoboken, New Jersey (2004)
10. Database of Human Iris. Institute of Automation of Chinese Academy of Sciences (CASIA). http://www.cbsr.ia.ac.cn/english/IrisDatabase.asp
11. Ma, L., Wang, Y., Tan, T.: Iris recognition based on multichannel Gabor filtering. In: Melbourne, Australia, ACCV2002. 5th Asian Conference on Computer Vision, vol. 1, pp. 279–283 (2002)
12. Muron, A., Pospisil, J.: The human iris structure and its usages. Czech Republic, Physica, pp. 89–95 (2000)
13. Mackey, M.C.: Adventures in Poland: having fun and doing research with Andrzej Lasota. Mat. Stosow 5–32 (2007)
14. Mackey, M.C., Glass, L.: Oscillation and chaos in physiological control systems. Science **197**, 287–289 (1997)
15. Melin, P., Castillo, O., Gonzalez, S., Cota, J., Trujillo, W., Osuna, P.: Design of Modular Neural Networks with Fuzzy Integration Applied to Time Series Prediction, vol. 41, pp. 265–273. Springer, Berlin (2007)
16. Dow Jones Company. http://www.dowjones.com (Jan 10, 2010)
17. Dow Jones Indexes. http://www.djindexes.com (Sept 5, 2010)
18. Castillo, O., Melin, P.: A review on the design and optimization of interval type-2 fuzzy controllers. Appl. Soft Comput. **12**(4), 1267–1278 (2012)
19. Melin, P.: Modular Neural Networks and Type-2 Fuzzy Systems for Pattern Recognition, pp. 1–204. Springer (2012)
20. Hagras, H.: Type-2 fuzzy logic controllers: a way forward for fuzzy systems in real world environments. In: IEEE World Congress on Computational Intelligence, pp. 181–200 (2008)
21. Sepúlveda, R., Castillo, O., Melin, P., Montiel, O.: An efficient computational method to implement type-2 fuzzy logic in control applications. Analysis and Design of Intelligent Systems using Soft Computing Techniques, pp. 45–52 (2007)
22. Castro, J., Castillo, O., Melin, P., Rodríguez-Díaz, A.: A hybrid learning algorithm for a class of interval type-2 fuzzy neural networks. Inf. Sci. **179**(13), 2175–2193 (2009)
23. Hidalgo, D., Melin, P., Castillo, O.: An optimization method for designing type-2 fuzzy inference systems based on the footprint of uncertainty using genetic algorithms. Expert Syst. Appl. **39**, 4590–4598 (2012)
24. Montiel, O., Castillo, O., Melin, P., Sepúlveda, R.: The evolutionary learning rule for system identification. Appl. Soft Comput. **3**(4), 343–352 (2003)

Chapter 4
Simulations and Results

This chapter shows simulation results of the optimization of the neural network ensemble with the genetic algorithm and PSO algorithm, as well as results for each of the type-1 and type-2 fuzzy integration and the optimization of these systems with GA and PSO. The main goal is finding the best network architecture for each of the time series: Mackey-Glass, Dow Jones and Mexican stock exchange and the integration with type-1 and type-2 fuzzy systems. We are aware that nothing can be sure in this world but at least we have to reduce the level of uncertainty in the forecast, which is an extremely important factor for some companies and organizations.

4.1 Iris Database

Five tests were performed with the proposed modular neural network under the same conditions and the same database of the iris; in Table 4.1 we show the obtained results.

The best result is a total recognition of 25 out of 30 images of iris of 10 persons; giving a recognition rate of 83.33 %.

The average of the 5 tests is 74.66 % of recognition.

In Table 4.1, we present a comparison of results with those of other researchers obtained for iris recognition of persons, realized with different or similar methods.

In Table 4.2, we present a comparison of results with those of other researchers obtained for iris recognition of persons, realized with different or similar methods.

© The Author(s) 2016
F. Gaxiola et al., *New Backpropagation Algorithm with Type-2 Fuzzy Weights for Neural Networks*, SpringerBriefs in Computational Intelligence, DOI 10.1007/978-3-319-34087-6_4

Table 4.1 Results from the experiments of recognition with the Iris database

No.	Epoch	Network error	Time (min)	Prediction error
E1	12	0.01	72	76.66 % (23/30)
E2	12	0.01	72	70 % (21/30)
E3	12	0.01	72	73.33 % (22/30)
E4	12	0.01	72	83.33 % (25/30)
E5	12	0.01	71	70 % (21/30)

Table 4.2 Table of comparison of results of recognition with the iris database

Method	Percentage of recognition (%)
Proposed method	83.33
Gaxiola	97.13
Sanchez-Avila	97.89
Ma	98.00
Tisse	89.37
Daugman	99.90
Sarhan	96.00

4.2 Mackey-Glass Time Series

This section presents the simulation and test results obtained by applying the proposed prediction method to the Mackey-Glass time series for $\tau = 17$, using different approach of the proposed method and the two types of optimization used in this work, the GA and PSO algorithm.

4.2.1 Ensemble Neural Network with Type-2 Fuzzy Weights for Mackey-Glass Time Series

The obtained results for the ensemble neural network with type-2 fuzzy weights are shown on Table 4.3 and Fig. 4.1. The best error is of 0.0788 without optimizing the neural network and type-2 fuzzy systems, which means that all parameters of the neural network and type-2 fuzzy systems are established empirically.

Table 4.3 Results for the ensemble neural network for series Mackey-Glass

No.	Epoch	Network error	Time	Prediction error
E1	100	0.000000001	00:01:09	0.0788
E2	100	0.000000001	00:02:11	0.0905
E3	100	0.000000001	00:02:12	0.0879
E4	100	0.000000001	00:01:14	0.0822
E5	100	0.000000001	00:01:13	0.0924
E6	100	0.000000001	00:02:13	0.0925
E7	100	0.000000001	00:01:08	0.0822
E8	100	0.000000001	00:01:09	0.0924
E9	100	0.000000001	00:01:07	0.0826
E10	100	0.000000001	00:01:07	0.0879

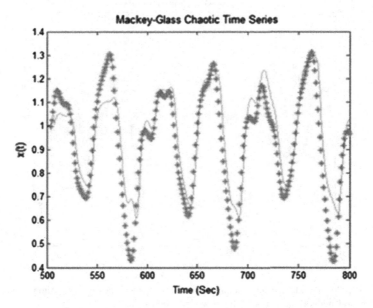

Fig. 4.1 Graphic of real data again prediction data of the ensemble neural network for Mackey-Glass time series

4.2.2 Optimization with Genetic Algorithm (GA) of the Ensemble Neural Network with Type-2 Fuzzy Weights for Mackey-Glass Time Series

The obtained results of the GA optimizing the ensemble neural network are shown on Table 4.4 and Fig. 4.2. The best error is of 0.0788 optimizing the numbers of neurons and type-2 fuzzy systems.

Table 4.4 Results for the
ensemble neural network for
series Mackey-Glass
optimized

No.	Epoch	Network error	Prediction error
E1	100	0.000000001	0.0518
E2	100	0.000000001	0.0611
E3	100	0.000000001	0.0787
E4	100	0.000000001	0.0715
E5	100	0.000000001	0.0655
E6	100	0.000000001	0.0614
E7	100	0.000000001	0.0724
E8	100	0.000000001	0.0712
E9	100	0.000000001	0.0724
E10	100	0.000000001	0.0518

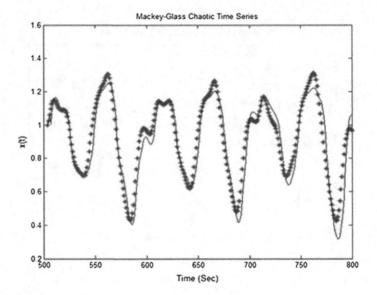

Fig. 4.2 Graphic of real data against prediction data of the Mackey-Glass time series for the ensemble neural network with type-2 fuzzy weights optimized

4.2.3   Interval Type-2 Fuzzy Weight Adjustment for Backpropagation Neural Networks with Application in Time Series Prediction

The obtained results for the experiments with the monolithic neural network are shown on Table 4.5 and Fig. 4.3, and all parameters of the neural network are established empirically. The best prediction error is of 0.055, and the average error is of 0.077.

Table 4.5 Results for the monolithic neural network in Mackey-Glass time series

No.	Epoch	Network error	Time (s)	Prediction error
E1	800	1×10^{-7}	06	0.091
E2	800	1×10^{-7}	06	0.088
E3	800	1×10^{-7}	06	0.055
E4	800	1×10^{-7}	06	0.083
E5	800	1×10^{-7}	05	0.080
E6	800	1×10^{-7}	05	0.077
E7	800	1×10^{-7}	06	0.094
E8	800	1×10^{-7}	06	0.071
E9	800	1×10^{-7}	06	0.055
E10	800	1×10^{-7}	06	0.066

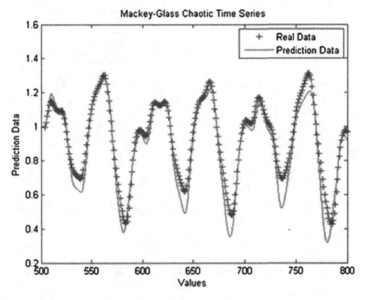

Fig. 4.3 Graphic of real data against prediction data of the Mackey-Glass time series for the monolithic neural network

We are presenting 10 experiments in Table 4.1, but the average error was calculated considering 40 experiments with the same parameters and conditions.

The obtained results for the experiments with the neural network with type-1 fuzzy weights are shown on Table 4.6 and Fig. 4.4; in this case all parameters of the neural network and type-1 fuzzy inference systems are established empirically. The best prediction error is of 0.055, and the average error is of 0.094.

The obtained results for the experiments with the neural network with type-2 fuzzy weights are shown on Table 4.7 and Fig. 4.5, in that all parameters of the neural network and type-2 fuzzy inference systems are established empirically.

Table 4.6 Results for the neural network with type-1 fuzzy weights (triangular MF) in Mackey-Glass time series

No.	Epoch	Network error	Time (s)	Prediction error
E1	100	1×10^{-8}	11	0.068
E2	100	1×10^{-8}	11	0.097
E3	100	1×10^{-8}	12	0.071
E4	100	1×10^{-8}	12	0.100
E5	100	1×10^{-8}	12	0.055
E6	100	1×10^{-8}	12	0.102
E7	100	1×10^{-8}	11	0.091
E8	100	1×10^{-8}	11	0.108
E9	100	1×10^{-8}	12	0.086
E10	100	1×10^{-8}	11	0.106

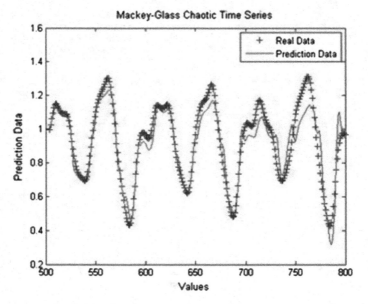

Fig. 4.4 Graphic of real data against prediction data of the Mackey-Glass time series for the neural network with type-1 fuzzy weights

We are also presenting results with different type-2 fuzzy inference systems; for this we vary the values of the triangular membership functions using a variable epsilon in different percentages. The best prediction error is of 0.039 (epsilon = ±80 %), and the best average error is of 0.061 (epsilon = ±80 %).

In Table 4.8 we are presenting the comparison among the monolithic neural network (MNN), the neural network with type-1 fuzzy weights (NNT1FW) and the best performance of the neural network with type-2 fuzzy weights (Eps = 80 %) (NNT2FW), which shows that the use of type-2 fuzzy weights has the best behavior.

Table 4.7 Results for the neural network with type-2 fuzzy weights for the Mackey-Glass time series

No.	Eps = 1 %	Eps = 5 %	Eps = 10 %	Eps = 20 %	Eps = 30 %	Eps = 40 %	Eps = 50 %	Eps = 60 %	Eps = 70 %	Eps = 80 %	Eps = 90 %
E1	0.085	0.086	0.065	0.063	0.060	0.068	0.081	0.076	0.075	0.063	0.110
E2	0.087	0.067	0.081	0.080	0.061	0.049	0.054	0.081	0.049	0.044	0.064
E3	0.072	0.077	0.090	0.060	0.064	0.050	0.077	0.083	0.044	0.058	0.084
E4	0.069	0.073	0.088	0.077	0.042	0.077	0.058	0.079	0.057	0.072	0.097
E5	0.054	0.060	0.059	0.049	0.058	0.053	0.079	0.058	0.073	0.047	0.115
E6	0.082	0.080	0.078	0.073	0.080	0.083	0.065	0.063	0.048	0.067	0.074
E7	0.045	0.072	0.086	0.067	0.072	0.063	0.069	0.050	0.078	0.049	0.123
E8	0.076	0.075	0.055	0.071	0.046	0.058	0.049	0.065	0.058	0.039	0.066
E9	0.079	0.069	0.075	0.069	0.083	0.082	0.072	0.053	0.069	0.070	0.090
E10	0.063	0.083	0.069	0.084	0.067	0.074	0.067	0.071	0.063	0.053	0.129
Average	0.085	0.077	0.078	0.076	0.071	0.072	0.072	0.073	0.068	0.061	0.098

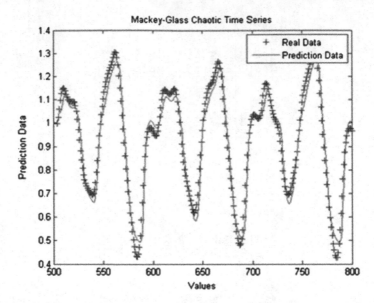

Fig. 4.5 Plot of real data against prediction data of the Mackey-Glass time series for the neural network with type-2 fuzzy weights (triangular MF) (epsilon = ±80 %)

Table 4.8 Comparison for the monolithic neural network, neural network with type-1 fuzzy weights and neural network with type-2 fuzzy weights (triangular MF) in time series Mackey-Glass

	Epoch	Network error	Time (s)	Prediction error
MNN	800	1×10^{-7}	06	0.053
NNT1FW	100	1×10^{-8}	12	0.055
NNT2FW	100	1×10^{-8}	44	0.039

We also performed an experiment applying noise in the range (0.1–1) in the test data to observe the behavior of the monolithic neural network, neural network with type-1 fuzzy weights and neural network with type-2 fuzzy weights. The obtained results for the experiments are shown on Table 4.5.

In Table 4.9, the row labeled MNN represents the results of the monolithic neural network, the row with Eps = 0 % represents the results of the neural network with type-1 fuzzy weights and the rows with Eps = 1 % to Eps = 90 % represent the results of the neural network with type-2 fuzzy weights for different levels of uncertainty. The best performance was achieved with the neural network with type-2 fuzzy weights (Eps = 50 %) with noise in the data test.

Table 4.9 Results for the monolithic neural network, neural network with type-1 fuzzy weights and neural network with type-2 fuzzy weights (triangular MF) in time series Mackey-Glass with noise (N)

No.	N = 0	N = 0.1	N = 0.2	N = 0.3	N = 0.4	N = 0.5	N = 0.6	N = 0.7	N = 0.8	N = 0.9	N = 1
MNN	0.053	1.098	1.157	1.328	1.279	1.309	1.280	1.282	1.309	1.366	1.344
Eps = 0 %	0.055	0.781	0.706	0.772	0.703	0.731	0.701	0.693	0.706	0.744	0.707
Eps = 1 %	0.045	0.339	0.359	0.450	0.432	0.464	0.453	0.469	0.483	0.507	0.491
Eps = 5 %	0.060	0.570	0.605	0.680	0.650	0.672	0.656	0.665	0.672	0.692	0.675
Eps = 10 %	0.055	0.563	0.552	0.621	0.578	0.606	0.590	0.596	0.602	0.624	0.604
Eps = 20 %	0.049	0.542	0.576	0.662	0.630	0.658	0.638	0.646	0.661	0.692	0.672
Eps = 30 %	0.042	0.702	0.741	0.824	0.762	0.768	0.728	0.723	0.738	0.755	0.725
Eps = 40 %	0.049	0.503	0.453	0.495	0.450	0.474	0.464	0.459	0.464	0.486	0.464
Eps = 50 %	0.049	0.281	0.261	0.287	0.257	0.273	0.266	0.269	0.267	0.272	0.256
Eps = 60 %	0.050	0.504	0.522	0.599	0.560	0.587	0.575	0.589	0.590	0.603	0.588
Eps = 70 %	0.044	0.529	0.546	0.585	0.549	0.565	0.560	0.580	0.570	0.558	0.553
Eps = 80 %	0.039	0.430	0.424	0.478	0.432	0.456	0.444	0.452	0.454	0.465	0.447
Eps = 90 %	0.064	1.302	1.365	1.479	1.474	1.497	1.471	1.485	1.512	1.545	1.536

4.2.4   Neural Network with Lower and Upper Type-2 Fuzzy Weights Using the Back-Propagation Learning Method

We performed experiments in time-series prediction, specifically for the Mackey-Glass time series ($\tau = 17$).

The obtained results of the experiments performed with the neural network in different percentage of epsilon for the type-2 fuzzy inference system are shown on Table 4.10 and Figs. 4.6, 4.7. The best prediction error is of 0.0321 and the average prediction error is of 0.0463 with epsilon = 0.8 %, these results are achieved without optimizing of the neural network and the type-2 fuzzy systems, which means that all parameters of the neural network and the range and values of the membership functions of the type-2 fuzzy systems are established empirically. The average error was obtained of 30 experiments.

In Table 4.10, we presents the prediction error obtained with the average of the lower and upper results achieved like output of the neural network.

In Fig. 4.6 we show the lower and upper prediction data with epsilon equal to 0.07 against the test data of the Mackey-Glass time series.

In Fig. 4.7 we show the lower and upper prediction data with epsilon equal to 0.1 against the test data of the Mackey-Glass time series.

4.2.5   Optimization of Type-2 Fuzzy Weight for Neural Network Using Genetic Algorithm and Particle Swarm Optimization

The obtained results with type-2 fuzzy inference systems without optimizing are shown on Table 4.11 and Fig. 4.8, which means that all parameters of the type-2 fuzzy systems are established empirically. The best error is of 0.0638.

Table 4.10 Prediction error for the neural network with interval type-2 fuzzy weights (lower and upper) for Mackey-Glass time series

Epsilon	0.01 %	0.07 %	0.1 %	0.2 %	0.8 %
1	0.0940	0.0678	0.0900	0.0600	0.0540
2	0.0840	0.0970	0.0790	0.0690	0.0321
3	0.1070	0.1070	0.0663	0.0890	0.0490
4	0.0772	0.0850	0.0820	0.0720	0.0420
5	0.0820	0.0900	0.0940	0.0627	0.0330
6	0.1150	0.1100	0.0730	0.0810	0.0460
7	0.0930	0.0760	0.0770	0.0790	0.0390
8	0.1010	0.1010	0.0920	0.0940	0.0560
9	0.1220	0.0860	0.0980	0.0490	0.0360
10	0.0960	0.0940	0.0880	0.0740	0.0440
Average error	0.1068	0.0935	0.0873	0.0805	0.0463

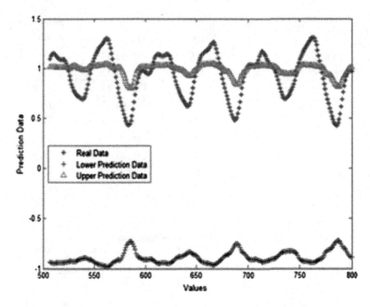

Fig. 4.6 Graphic of the lower and upper prediction data with epsilon = 0.07 % against the test data of Mackey-Glass time series

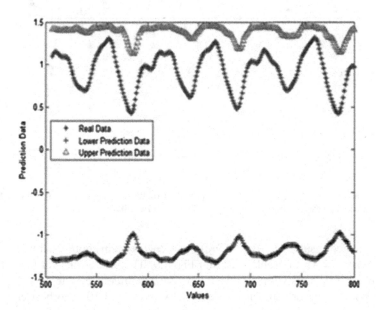

Fig. 4.7 Graphic of the lower and upper prediction data with epsilon = 0.1 % against the test data of Mackey-Glass time series

No.	Epoch	Time	Prediction error
E1	100	00:01:09	0.0692
E2	100	00:01:11	0.0670
E3	100	00:01:12	0.0745
E4	100	00:01:14	0.0638
E5	100	00:01:13	0.0719
E6	100	00:01:13	0.0659
E7	100	00:01:08	0.0784
E8	100	00:01:09	0.0819
E9	100	00:01:07	0.0795
E10	100	00:01:07	0.0730

Table 4.11 Results for the neural network with type-2 fuzzy weights for NNT2FWGA-PSO in Mackey-Glass time series

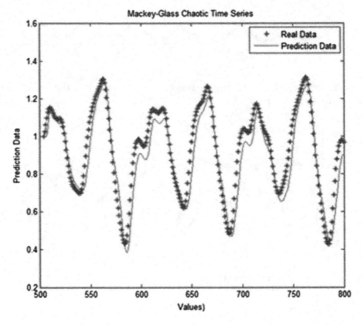

Fig. 4.8 Graphic of real data against prediction data of the Mackey-Glass time series for the neural network with type-2 fuzzy weights for NNT2FWGA-PSO

The obtained results with type-2 fuzzy inference systems optimized with genetic algorithm are shown on Table 4.12 and Fig. 4.9. The best error is of 0.0411.

The obtained results with optimized type-2 fuzzy inference systems with particle swarm optimization are shown on Table 4.13 and Fig. 4.10. The best error is of 0.0506.

The results for the three experiments are shown on Table 4.14.

Table 4.12 Results for the neural network using type-2 fuzzy inference systems optimized with genetic algorithm for Mackey-Glass time series

No.	Generations	Individuals (Real)	Prediction error
E1	65	100	0.0502
E2	65	100	0.0526
E3	65	100	0.0501
E4	65	100	0.0431
E5	65	100	0.0522
E6	65	100	0.0411
E7	65	100	0.0553
E8	65	100	0.0455
E9	65	100	0.0454
E10	65	100	0.0431

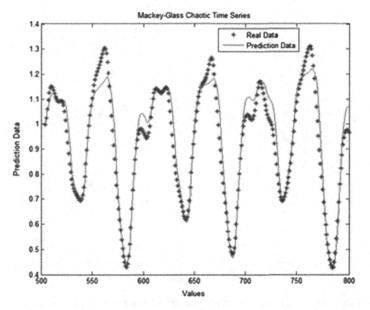

Fig. 4.9 Graphic of real data against prediction data of the Mackey-Glass time series for the neural network with type-2 fuzzy weights optimized with genetic algorithms

4.2.6 Neural Network with Fuzzy Weights Using Type-1 and Type-2 Fuzzy Learning with Gaussian Membership Functions

We have presented the results of the experiments performed with the neural network with type-1 fuzzy weights (NNT1FW) and the neural network with type-1 fuzzy weights (NNT1FW), these results are achieved without optimizing of the neural network and the type-1 fuzzy systems, which means that all parameters of

Table 4.13 Results for the neural network using type-2 fuzzy inference systems optimized with particle swarm optimization for Mackey-Glass time series

No.	Iterations	Particles (Real)	Prediction error
E1	65	100	0.0668
E2	65	100	0.0544
E3	65	100	0.0582
E4	65	100	0.0506
E5	65	100	0.0647
E6	65	100	0.0640
E7	65	100	0.0558
E8	65	100	0.0809
E9	65	100	0.0708
E10	65	100	0.0596

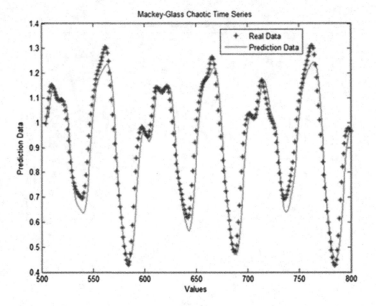

Fig. 4.10 Plot of real data against prediction data of the Mackey-Glass time series for the neural network with type-2 fuzzy weights optimized with particle swarm optimization

Table 4.14 Results for the three experiments for Mackey-Glass time series for NNT2FWGA-PSO

Experiment	Prediction error
Neural network with type-2 fuzzy inference systems without optimization	0.0638
NN with type-2 fuzzy inference systems optimizing with genetic algorithm	0.0411
NN with type-2 fuzzy inference systems optimizing with particle swarm optimization	0.0506

Table 4.15 Prediction error for the neural network with type-1 fuzzy weights (Gaussian MF) for Mackey-Glass time series

No.	Epoch	Network error	Prediction error
E1	100	1×10^{-8}	0.0704
E2	100	1×10^{-8}	0.0554
E3	100	1×10^{-8}	0.0652
E4	100	1×10^{-8}	0.0489
E5	100	1×10^{-8}	0.0601
E6	100	1×10^{-8}	0.0615
E7	100	1×10^{-8}	0.0684
E8	100	1×10^{-8}	0.0751
E9	100	1×10^{-8}	0.0737
E10	100	1×10^{-8}	0.0711
Average prediction error			0.0668

the neural network and the range and values of the membership functions of the type-1 fuzzy systems are established empirically. The average error was obtained of 30 experiments.

In Table 4.15, we present the prediction error obtained with the results achieved as output of NNT1FW. The best prediction error is of 0.0489 and the average prediction error is of 0.0668.

In Fig. 4.11 we show prediction data with type-1 fuzzy weights against the test data of the Mackey-Glass time series.

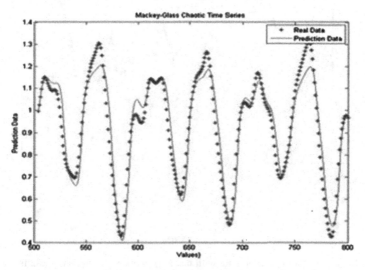

Fig. 4.11 Graphic of the prediction data with Neural Network with type-1 fuzzy weights (Gaussian MF) against the test data of the Mackey-Glass time series

In Table 4.16, we present the prediction error obtained with the results achieved as output of NNT2FW. The best prediction error is of 0.0413 and the average prediction error is of 0.0579.

In Fig. 4.12 we show the prediction data with type-2 fuzzy weights against the test data of the Mackey-Glass time series.

Table 4.16 Prediction error for the neural network with interval type-2 fuzzy weights (Gaussian MF) for the Mackey-Glass time series

No.	Epoch	Network error	Prediction error
E1	100	1×10^{-8}	0.0549
E2	100	1×10^{-8}	0.0691
E3	100	1×10^{-8}	0.0557
E4	100	1×10^{-8}	0.0432
E5	100	1×10^{-8}	0.0602
E6	100	1×10^{-8}	0.0673
E7	100	1×10^{-8}	0.0413
E8	100	1×10^{-8}	0.0618
E9	100	1×10^{-8}	0.0645
E10	100	1×10^{-8}	0.0522
Average prediction error			0.0579

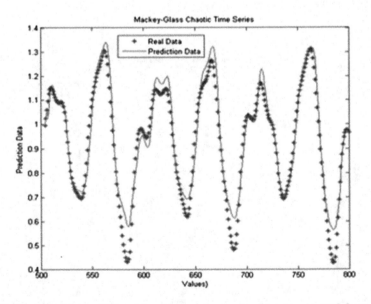

Fig. 4.12 Graphic of the prediction data with neural network with interval type-2 fuzzy weights (Gaussian MF) against the test data of the Mackey-Glass time series

4.2.7 Comparison of Neural Network Performance Under Different Membership Functions in the Type-2 Fuzzy Weights (CMFNNT2FW)

We present the obtained results of the experiments performed with the neural network with type-2 fuzzy weights (NNT2FW) in the different types: triangular, trapezoidal, two-sided Gaussian, and generalized bell; These results are achieved without optimizing of the neural network and the type-2 fuzzy systems, which means that all parameters of the neural network and the range and values of the membership functions of the type-2 fuzzy systems are established empirically. The average error was obtained of 30 experiments.

In Table 4.17, we present the prediction error obtained with the results achieved as output. The best prediction error is of 0.0391 and the average prediction error is of 0.0789. The APE represents the average prediction error of the experiments.

In Fig. 4.13 we show the best prediction data with the type-2 fuzzy weights against the test data of the Mackey-Glass time series.

4.2.8 Comparison of Results for the Mackey-Glass Time Series Neural Network with Fuzzy Weights

In Table 4.18, the results of the best prediction error for the proposed method of the neural network with fuzzy weights are shown.

Table 4.17 Prediction error for the neural network with type-2 fuzzy weights for Mackey-Glass time series, with the different type and variants of membership functions

Experiment	Triangular	Trapezoidal	Gaussian	Generalized bell
1	0.0768	0.0892	0.0391	0.0947
2	0.0658	0.1133	0.0779	0.1454
3	0.0719	0.1195	0.0565	0.1772
4	0.0552	0.0799	0.0718	0.1165
5	0.0728	0.1273	0.0355	0.1870
6	0.0661	0.1070	0.0688	0.1356
7	0.0808	0.1511	0.0796	0.1586
8	0.0686	0.1320	0.0589	0.1156
9	0.0888	0.0927	0.0454	0.1642
10	0.0866	0.1490	0.0693	0.1275
APE	0.0789	0.1272	0.0659	0.1574

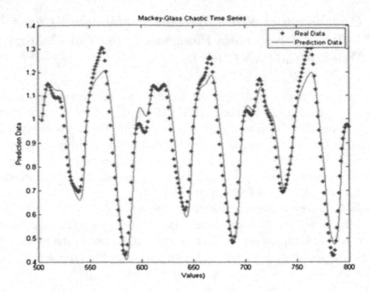

Fig. 4.13 Graphic of the best prediction data with NNT2FW with the different type and variants of membership functions against the test data of the Mackey-Glass time series

Table 4.18 Results for the all experiments for Mackey-Glass time series for neural network with fuzzy weights

Experiment	Prediction error
Monolithic neural network	0.0530
Neural network (NN) with type-1 fuzzy inference systems (triangular MF)	0.0550
NN with type-1 fuzzy inference systems (Gaussian MF)	0.0489
Ensemble NN with type-2 fuzzy weights without optimization (triangular MF)	0.0788
Ensemble NN with type-2 fuzzy weights optimized with GA (triangular MF)	0.0518
NN with type-2 fuzzy inference systems Eps = 80 % without optimization (triangular MF)	0.0390
NN with type-2 fuzzy inference systems without optimization (Gaussian MF)	0.0355
NN with type-2 fuzzy inference systems without optimization (triangular MF)	0.0638
NN with type-2 fuzzy inference systems optimizing with genetic algorithm (triangular MF)	0.0411
NN with type-2 fuzzy inference systems optimizing with particle swarm optimization (triangular MF)	0.0506

4.3 Dow-Jones Time Series

This section presents the simulation and test results obtained by applying the proposed prediction method to the Dow-Jones time series.

Neural network with fuzzy weights using type-1 and type-2 fuzzy learning with Gaussian membership functions

Table 4.19 Prediction error for the neural network with type-1 fuzzy weights for Dow-Jones time series

No.	Epoch	Network error	Prediction error
E1	100	1×10^{-8}	0.0121
E2	100	1×10^{-8}	0.0195
E3	100	1×10^{-8}	0.0198
E4	100	1×10^{-8}	0.0097
E5	100	1×10^{-8}	0.0238
E6	100	1×10^{-8}	0.0213
E7	100	1×10^{-8}	0.0231
E8	100	1×10^{-8}	0.0171
E9	100	1×10^{-8}	0.0216
E10	100	1×10^{-8}	0.0147
Average prediction error			0.0201

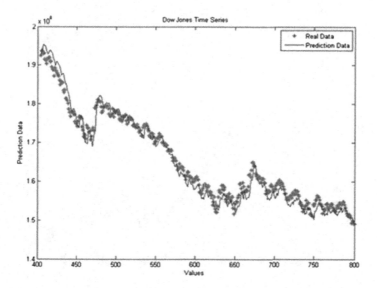

Fig. 4.14 Plot of the prediction data with NNT1FW against the test data of the Dow-Jones time series

We presented the obtained results of the experiments performed with the neural network with type-1 fuzzy weights (NNT1FW) and the neural network with type-1 fuzzy weights (NNT1FW), these results are achieved without optimizing of the neural network and the type-1 fuzzy systems, which means that all parameters of the neural network and the range and values of the membership functions of the type-1 fuzzy systems are established empirically. The average error was obtained of 30 experiments.

Table 4.20 Prediction error for the neural network with interval type-2 fuzzy weights for the Dow-Jones time series

No.	Epoch	Network error	Prediction error
E1	100	1×10^{-8}	0.0105
E2	100	1×10^{-8}	0.0082
E3	100	1×10^{-8}	0.0125
E4	100	1×10^{-8}	0.0114
E5	100	1×10^{-8}	0.0108
E6	100	1×10^{-8}	0.0095
E7	100	1×10^{-8}	0.0080
E8	100	1×10^{-8}	0.0101
E9	100	1×10^{-8}	0.0093
E10	100	1×10^{-8}	0.0117
Average prediction error			0.0104

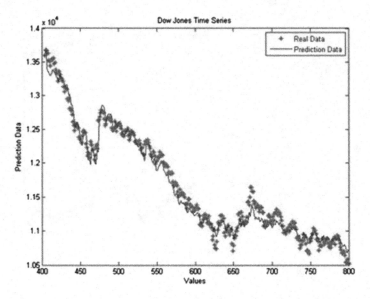

Fig. 4.15 Plot of the prediction data with NNT2FW against the test data of the Dow-Jones time series

In Table 4.19, we present the prediction error obtained with the results achieved as output of NNT1FW. The best prediction error is of 0.0097 and the average prediction error is of 0.0201.

In Fig. 4.14 we show prediction data with type-1 fuzzy weights against the test data of the Dow-Jones time series.

In Table 4.20, we present the prediction error obtained with the results achieved as output of NNT2FW. The best prediction error is of 0.0080 and the average prediction error is of 0.0104.

In Fig. 4.15 we show the prediction data with type-2 fuzzy weights against the test data of the Dow-Jones time series.

Chapter 5
Conclusions

Upon completion of this book the effectiveness of the neural networks with type-2 fuzzy weights for forecasting time series has been proven as it allows the problem can be solved and thus obtained satisfactory results, specifically for Mackey-Glass time series. As was the case in this book, the results obtained in each of the series that are considered good because the margin of error that was obtained is very small.

The neural network with Gaussian type-2 fuzzy weights shows better behavior at different levels of uncertainty than the monolithic neural network and neural network with type-1 fuzzy weights, this conclusion is based on the fact that best values of Table 4.18 are with Gaussian type-2 fuzzy weights. The results obtained for the prediction error obtained are smaller than the results of the others methods.

Applying different levels of noise we observe that the neural network with type-2 fuzzy weights has better behavior and tolerance to noise than monolithic neural network and neural network with type-1 fuzzy weights. This conclusion was found by observing that the type-2 fuzzy weights present lower prediction errors than the other methods.

The results obtained in these experiments show that the neural network with type-2 fuzzy weights obtained better results, without noise and with noise, than the monolithic neural network and the neural network with type-1 fuzzy weights, in the prediction for the Mackey-Glass time series.

The proposed methods, neural network with type-1 and type-2 fuzzy weights, have more robustness and achieve better results than the monolithic neural network. Besides, the type-2 fuzzy weights provide the neural network less susceptibility to the occurrence of a significant increase in the prediction error when noise is applied to the real data.

Another Objective was to optimize the type-2 fuzzy inference systems used in the neural network to obtain the fuzzy weights, so we work with genetic algorithm and particle swarm optimization for optimize the triangular type-2 fuzzy inference systems, and these presents' better results than the un-optimized type-2 fuzzy inference systems.

© The Author(s) 2016   99
F. Gaxiola et al., *New Backpropagation Algorithm with Type-2 Fuzzy Weights
for Neural Networks*, SpringerBriefs in Computational Intelligence,
DOI 10.1007/978-3-319-34087-6_5

In persons recognition using the iris biometric measure, the proposed method not obtain better results than the other methods; however, we presented only one experiment for this area, so we need more work and experiments for prove that the proposed method is good for the person recognition [1].

Reference

1. Arellano, M.: Introduction to time series analysis (2001).

Index

© The Author(s) 2016   101
F. Gaxiola et al., *New Backpropagation Algorithm with Type-2 Fuzzy Weights
for Neural Networks*, SpringerBriefs in Computational Intelligence,
DOI 10.1007/978-3-319-34087-6

Printed in the United States
By Bookmasters

Printed in the United States
By Bookmasters